High Performance Computing and the Discrete Element Model

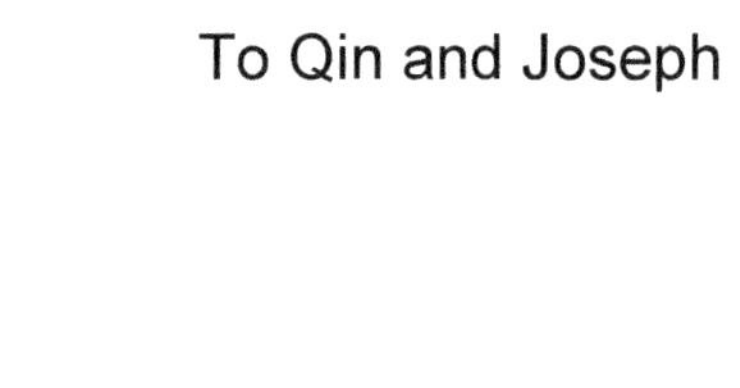
To Qin and Joseph

Discrete Granular Mechanics Set

coordinated by
Félix Darve

High Performance Computing and the Discrete Element Model

Opportunity and Challenge

Gao-Feng Zhao

First published 2015 in Great Britain and the United States by ISTE Press Ltd and Elsevier Ltd

Apart from any fair dealing for the purposes of research or private study, or criticism or review, as permitted under the Copyright, Designs and Patents Act 1988, this publication may only be reproduced, stored or transmitted, in any form or by any means, with the prior permission in writing of the publishers, or in the case of reprographic reproduction in accordance with the terms and licenses issued by the CLA. Enquiries concerning reproduction outside these terms should be sent to the publishers at the undermentioned address:

ISTE Press Ltd
27-37 St George's Road
London SW19 4EU
UK

www.iste.co.uk

Elsevier Ltd
The Boulevard, Langford Lane
Kidlington, Oxford, OX5 1GB
UK

www.elsevier.com

Notices
Knowledge and best practice in this field are constantly changing. As new research and experience broaden our understanding, changes in research methods, professional practices, or medical treatment may become necessary.

Practitioners and researchers must always rely on their own experience and knowledge in evaluating and using any information, methods, compounds, or experiments described herein. In using such information or methods they should be mindful of their own safety and the safety of others, including parties for whom they have a professional responsibility.

To the fullest extent of the law, neither the Publisher nor the authors, contributors, or editors, assume any liability for any injury and/or damage to persons or property as a matter of products liability, negligence or otherwise, or from any use or operation of any methods, products, instructions, or ideas contained in the material herein.

MATLAB® is a trademark of The MathWorks, Inc. and is used with permission. The MathWorks does not warrant the accuracy of the text or exercises in this book. This book's use or discussion of MATLAB® software or related products does not constitute endorsement or sponsorship by The MathWorks of a particular pedagogical approach or particular use of the MATLAB® software.

For information on all our publications visit our website at http://store.elsevier.com/

© ISTE Press Ltd 2015
The rights of Gao-Feng Zhao to be identified as the author of this work have been asserted by him in accordance with the Copyright, Designs and Patents Act 1988.

British Library Cataloguing-in-Publication Data
A CIP record for this book is available from the British Library
Library of Congress Cataloging in Publication Data
A catalog record for this book is available from the Library of Congress
ISBN 978-1-78548-031-7

Printed and bound in the UK and US

Contents

Foreword

Molecular dynamics is recognized as a powerful method in modern computational physics. This method is essentially based on a factual observation: the apparent strong complexity and extreme variety of natural phenomena are not due to the intrinsic complexity of the element laws but due to the very large number of basic elements in interaction through, in fact, simple laws. This is particularly true for granular materials in which a single intergranular friction coefficient between rigid grains is enough to simulate, at a macroscopic scale, the very intricate behavior of sand with a Mohr–Coulomb plasticity criterion, a dilatant behavior under shearing, non-associate plastic strains, etc. and, in fine, an incrementally nonlinear constitutive relation. Passing in a natural way from the grain scale to the sample scale, the discrete element method (DEM) is precisely able to bridge the gap between micro- and macro-scales in a very realistic way, as it is today verified in many mechanics labs.

Thus, DEM is today in an impetuous development in geomechanics and in the other scientific and technical fields related to grain manipulation. Here lies the basic reason for this new set of books called "Discrete Granular Mechanics", in which not only numerical questions are considered but also experimental, theoretical and analytical aspects in relation to the discrete nature of granular media. Indeed, from an experimental point of view, computational tomography – for example – is giving rise today to the description of all the translations and rotations of a few thousand grains inside a given sample and to the identification of the formation of mesostructures such as force chains and force loops. With respect to theoretical aspects, DEM is also confirming, informing or at least precising some theoretical clues such as the questions of failure modes, of the expression of stresses inside a partially saturated medium and of the mechanisms involved in granular avalanches. Effectively, this set has been planned to cover all the experimental, theoretical and numerical approaches related to discrete granular mechanics.

The observations show undoubtedly that granular materials have a double nature, that is continuous and discrete. Indeed, roughly speaking, these media respect the matter continuity at a macroscopic scale, whereas they are essentially discrete at the granular microscopic scale. However, it appears that, even at the macroscopic scale, the discrete aspect is still present. An emblematic example is constituted by the question of shear band thickness. In the framework of continuum mechanics, it is well recognized that this thickness can be obtained only by introducing a so-called "internal length" through "enriched" continua. However, this internal length seems to be not intrinsic and to constitute a kind a constitutive relation by itself. Probably, it is because to consider the discrete nature of the medium by a simple scalar is oversimplifying reality. However, in a DEM modeling, this thickness is obtained in a natural way without any *ad hoc* assumption. Another point, whose proper description was indomitable in a continuum mechanics approach, is the post-failure behavior. The finite element method, which is essentially based on the inversion of a stiffness matrix, which is becoming singular at a failure state, meets some numerical difficulties to go beyond a failure state. Here also, it appears that DEM is able to simulate fragile, ductile, localized or diffuse failure modes in a direct and realistic way – even in some extreme cases such as fragmentation rupture.

The main limitation of DEM is probably linked today to the limited number of grains or particles, which can be considered in relation to an acceptable computation time. Thus, the simulation of boundary value problems stays, in fact, bounded by more or less heuristic cases. So, the current computations in labs involve at best a few hundred thousand grains and, for specific problems, a few million. Let us note however that the parallelization of DEM codes has given rise to some computations involving 10 billion grains, thus opening widely the field of applications for the future.

In addition, this set of books will also present the recent developments occurring in micromechanics, applied to granular assemblies. The classical schemes consider a representative element volume. These schemes are proposing to go from the macro-strain to the displacement field by a localization operator, then the local intergranular law relates the incremental force field to this incremental displacement field, and eventually a homogenization operator deduces the macro-stress tensor from this force field. The other possibility is to pass from the macro-stress to the macro-strain by considering a reverse path. So, some macroscopic constitutive relations can be established, which properly consider an intergranular incremental law. The greatest advantage of these micromechanical relations is probably to consider only a few material parameters, each one with a clear physical meaning.

This set of around 20 books has been envisaged as an overview toward all the promising future developments mentioned earlier.

Félix Darve
July 2015

Preface

The classical continuum theory can adequately describe the macroscopic mechanical response of most artificial materials. However, this theory breaks down when facing critical problems in geomechanics (e.g. progressive failure and strain localization) because of the discontinuous nature of rock masses and granular soils. To tackle these problems, discontinuum-based models, for example the discrete element model (DEM), have been developed. The DEM has been increasingly applied to many geomechanical problems encountered in civil, mining, hydropower and petroleum engineering. Despite many advantages, the high computational requirement is one of the main drawbacks of this method. Modern computers have provided powerful hardware platforms for high performance computing; however, existing DEM codes are usually programmed serially, hindering the ability to fully use modern computing resources. A parallel DEM code is therefore needed. This book describes the parallel implementation of a DEM code and covers a wide scope for DEM (from algorithms to modeling techniques and engineering applications). This book will be a valuable reference for researchers interested in geomechanics and discontinuum-based models.

Gao-Feng Zhao
July 2015

Introduction

The discrete element model (DEM) has become a popular numerical tool in both scientific research and engineering applications. Despite many advantages, high computational requirement is one of the primary drawbacks of the method [CUN 01]. Modern computers have provided powerful hardware platforms for High Performance Computing (HPC); however, existing DEM codes are usually programmed in a serial manner, making them unable to use modern computing resources fully. Thus, a parallel DEM code is usually required.

I.1. Discrete element model

The DEM was first introduced by Cundall [CUN 71] to solve problems in rock mechanics. After being developed for nearly half a century, it has been applied to much broader areas, for example granular flow [WAL 09], fracturing of rock [LIS 14], unsaturated soil [LIU 03], chemical engineering [ZHU 08] and self-assembly [FAN 11]. The reasons for its popularity are as follows: (1) the simulation is closer to the physical world compared to the continuum mechanics-based models; and (2) the underlying principle is straightforward and easily understood. The methodology of the DEM can be simply interpreted as representing the behavior of matter through interactions of an assembly of rigid or deformable blocks/particles/bodies under given governing physical laws. For example, the calculation core of the widely used DEM for mechanical analysis is shown in Figure I.1.

As shown in the figure, given particle displacements (either prescribed initially or obtained from the previous time step), particle forces are calculated according to the corresponding constitutive models. Then, the particle's motion state (velocity, acceleration and position) is updated according to Newton's second law of motion.

Then, the velocities are updated as follows:

$$\dot{u}_i^{(t+\Delta t/2)} = \dot{u}_i^{(t-\Delta t/2)} + \frac{\sum F_j^{(t)}}{m_p} \Delta t \quad \text{[I.1]}$$

where $\dot{u}_i^{(t+\Delta t/2)}$ and $\dot{u}_i^{(t-\Delta t/2)}$ represent the particle velocities at $t+\Delta t/2$ and $t-\Delta t/2$, respectively; $\sum F_j^{(t)}$ is the total force applied to the particle *i*; m_p is the particle mass and Δt is the time step.

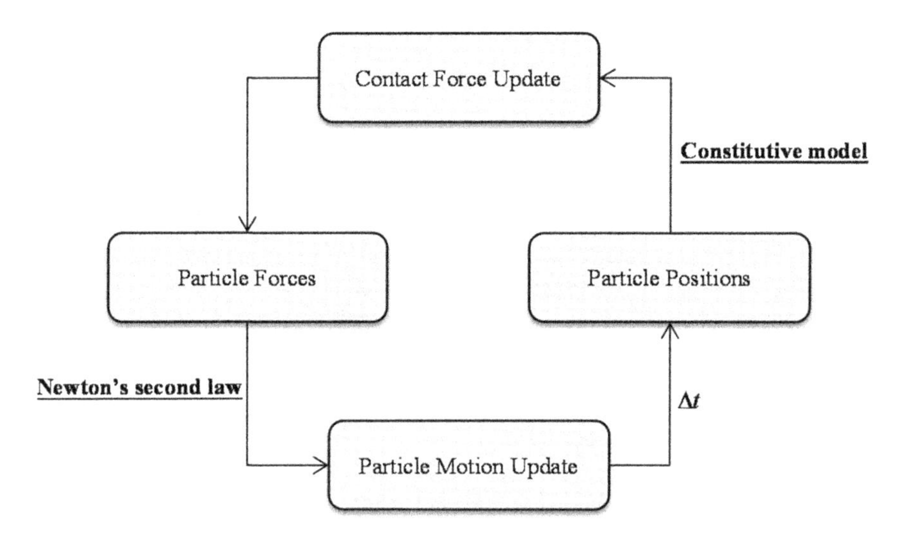

Figure I.1. *Calculation cycle of the DEM for mechanical analysis*

The particle displacements are obtained as follows:

$$\mathbf{u}_i^{(t+\Delta t)} = \mathbf{u}_i^{(t)} + \dot{\mathbf{u}}_i^{(t+\Delta t/2)} \Delta t \quad \text{[I.2]}$$

where $\mathbf{u}_i^{(t+\Delta t)}$ and $\mathbf{u}_i^{(t)}$ represent the displacement at $t + \Delta t$ and *t*, respectively.

To simplify the implementation, MATLAB® was selected as the programming environment in this book. A 45-line DEM code for Galileo's Leaning Tower of Pisa experiment is developed using equations [I.1] and [I.2]. As shown in Figure I.2, the 45-line DEM code is made from a preprocessor, a solver and a postprocessor. The preprocessor is required to (1) build a particle model; (2) assign material parameters; (3) provide initial condition; (4) apply loading conditions and (5) set the simulation time, time step, etc. The solver performs iterations of the DEM calculation cycle, as shown in Figure I.1. It is usually the most costly computational component of a DEM code.

```
function HelloDEM % Galileo's Leaning Tower of Pisa experiment
M_Ball_Steel=10;%10kg steel ball
M_Ball_Wood=1;% 1kg wood ball
Y0_Ball_Steel=100;%Initial position of the steel ball
Y0_Ball_Wood=100;%Initial position of the steel ball
dT=0.01;%Time step
NLoops=500;%Number of cycles
G=9.8;%Gravitational constant                              Preprocessor
U_Ball_Steel=0;%Initial displacement etc.
U_Ball_Wood=0;
V_Ball_Steel=0;
V_Ball_Wood=0;
A_Ball_Steel=0;
A_Ball_Wood=0;
Y_Ball_Steel=Y0_Ball_Steel;%Initial position of the steel ball
Y_Ball_Wood=Y0_Ball_Wood;%Initial position of the steel ball
t=0;
t_H=[];%History of time etc.
Y_H_Ball_Steel=[];
Y_H_Ball_Wood=[];
for i=1:NLoops
  F_Ball_Steel=-G*M_Ball_Steel; %Particle forces       Constitutive model
  F_Ball_Wood=-G*M_Ball_Wood;
  A_Ball_Steel=F_Ball_Steel/M_Ball_Steel; %The Newton's second law
  A_Ball_Wood=F_Ball_Wood/M_Ball_Wood;
  V_Ball_Steel=V_Ball_Steel+A_Ball_Steel*dT; %Update the velocity
  V_Ball_Wood=V_Ball_Wood+A_Ball_Wood*dT;
  U_Ball_Steel=U_Ball_Steel+V_Ball_Steel*dT; %Update the displacement
  U_Ball_Wood=U_Ball_Wood+V_Ball_Wood*dT;                   Solver
  Y_Ball_Steel=Y0_Ball_Steel+U_Ball_Steel;%Update the position
  Y_Ball_Wood=Y0_Ball_Wood+U_Ball_Wood;
  t=t+dT;
  %Record the history
  t_H=[t_H,t];
  Y_H_Ball_Steel=[Y_H_Ball_Steel,Y_Ball_Steel];
  Y_H_Ball_Wood=[Y_H_Ball_Wood,Y_Ball_Wood];
end
plot(t_H,Y_H_Ball_Steel,'k','linewidth',2);
hold on;
plot(t_H,Y_H_Ball_Steel,'g--','linewidth',1);
Ground=zeros(1,length(t_H));
plot(t_H,Ground,'r','linewidth',2);                       Postprocessor
xlabel('Time (s)')
ylabel('Height (m)');
legend('Steel Ball','Wood Ball','Ground');
```

Figure I.2. *A 45-line DEM code for Galileo's Leaning Tower of Pisa experiment*

Many researchers consider contact detection and contact treatment to be the most distinct parts of a DEM code. However, the author believes that the constitutive model of a DEM code is its spirit. First, the constitutive model covers contact detection (e.g. when there are interactions between two particles) and contact treatment (e.g. how should interaction forces between two particles be applied and distributed?). Moreover, to address different problems, specific constitutive models must be developed and implemented into the DEM. For example, a long-range magnetic force model is needed for a magnetic self-assembling problem, a Coulomb friction model is needed for granular flow simulation and a bond constitutive model is necessary for the dynamic fracturing of rock. Therefore, the constitutive model is the key component of a DEM solver. In actual application, the preprocessor is also very important. From the point of view of a DEM user, the preprocessor is the interface from the real world to the digital world. It transfers the user's abstract ideas or concepts into numerical models in the computer. In addition to the preprocessor, the postprocessor is also a crucial component of a DEM code. It can be regarded as the interface from the digital world to the real world. The results of DEM simulation usually comprise massive data, for example particle positions and particle velocities. To extract essential and easily understood information from the numerical results, the user requires a postprocessor. For example, the postprocessor of the 45-line DEM code plots the position histories of a steel ball and wood ball together with the ground profile (Figure I.3). From the figure, the times when the two balls hit the ground can be directly observed. The same conclusion that Galileo reached in 1589 can be easily achieved.

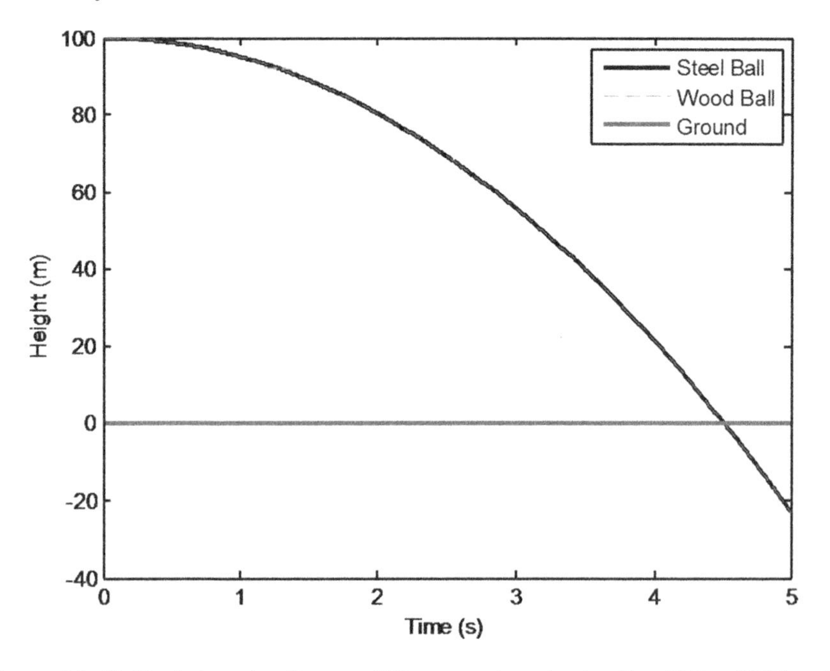

Figure I.3. *Galileo's Leaning Tower of Pisa experiment using the 45-line DEM code*

The 45-line DEM code is only a simple demonstrative example. To solve actual problems in scientific research and engineering applications, the researchers need to extend the code to handle thousands or even millions of particles governed by much more complex constitutive models. This results in the cost of the actual DEM code being computationally very high. Fortunately, modern computers provide HPC solutions to tackle this problem.

I.2. Discrete element model and high performance computing

With the development of computer technology, the supercomputer of the past is no longer *super*. Figure I.4 shows the computing performances of the world's top 500 supercomputers and the author's laptop (ThinkPad W540).

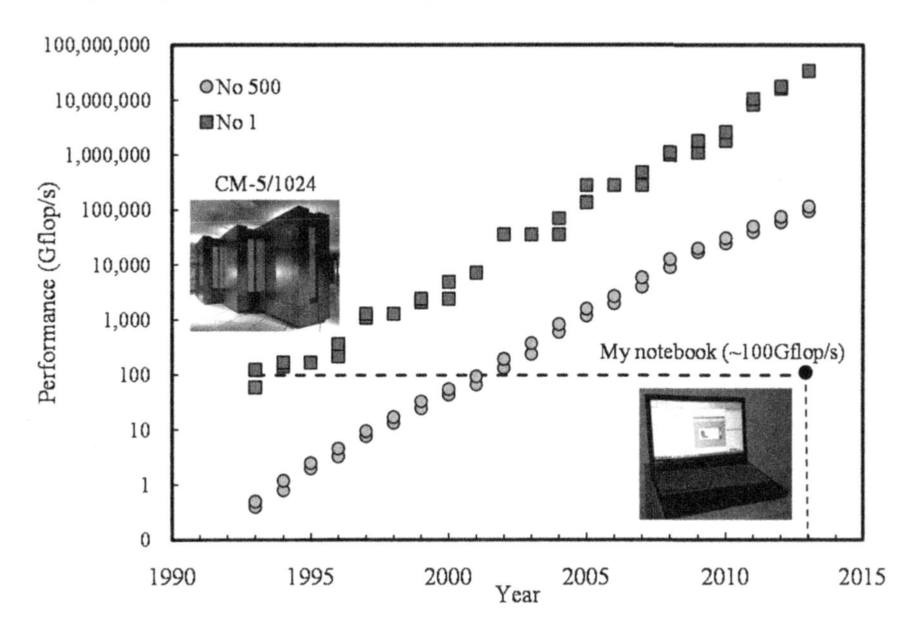

Figure I.4. *Top 500 supercomputers and the author's laptop (data obtained from http://www.top500.org/ on 10 July 2014)*

The laptop is found to be even faster than the world's most powerful supercomputer in 1992. However, running a serial DEM code on the laptop cannot be considered as HPC. Therefore, in this book, HPC refers to parallel computing rather than to its literal meaning. Here, HPC is synonymous with parallel computing, which distributes a single computational task to several processors and executes the distributed works simultaneously.

Implementation of a parallel DEM code used to be a complex task; fortunately, with the development of hardware and software technologies in computer science, it is now much easier. Currently, three choices for parallelization of a code are available: the multi-core computer, graphics processing unit (GPU) computer and cluster. The multi-core computer is a personal computer (PC) that is equipped with multiple processors and uses a shared-memory configuration. In addition to a specially designed shared-memory supercomputer, modern PCs are usually typical multi-core computers. For example, the author's laptop is equipped with a central processing unit (CPU) of four processors and has 8 GB physical memory. The GPU computer is another economic choice for HPC. GPU computing is also called heterogeneous or GPU/CPU coupled computing. Its basic principle is to calculate mathematically intensive tasks using a specially designed GPU card rather than a CPU. The third choice is a high-level parallelization system called cluster, which usually comprises many computer nodes (each node can be a multi-core PC, a GPU computer or even a PlayStation). At present, cluster is still the only feasible choice for very massive computing (e.g. more than billions of particles). In addition to hardware platforms, the software programming environment is also an important factor that has to be considered in the parallel implementation of a DEM code. The most commonly used parallel techniques are the message passing interface (MPI), Open Multi-Processing (OpenMP), compute unified device architecture (CUDA) and Open Computing Language (OpenCL). Until now, parallel DEM codes have been developed in various hardware platforms using these software programming environments and have achieved considerable speedups (e.g. see Table I.1).

Literature	Hardware	Software	Speedup
[DOW 99]	Cluster (CM5, 64 processors)	MPI	60
[ABC 04]	Cluster (SGI Origin 3800, 108 processors)	MPI	~80
[SCH 04]	PC with FPGA cards	OpenCL	~60
[MAK 06]	Cluster (VILKAS, 16 processors)	MPI	~11
[ZSA 09]	Shared-memory supercomputer (SGI Origin 2000, 8 processors)	OpenMP	~7
[KAC 10]	Cluster (VILKAS, 10 processors)	MPI	9
[MAR 11]	Cluster (VILKAS, 48 processors)	MPI	~32
[NIS 11]	Shared-memory supercomputer (Altix 4700, 256 cores)	OpenMP	~25.6
[ZHA 12]	PC (NVIDIA GeForce GTX 580)	CUDA	23
[ZHE 12]	PC (GTX 580)	CUDA	29
[ZHA 13b]	Cluster (Pleiades 2, 256 processors)	MPI	41
[GOP 13]	Cluster (256 processors)	MPI	~198
[WAN 13]	PC (NVIDIA GTX 580)	CUDA	417
[ZHA 13a]	PC (NVIDIA GTS 250)	CUDA	147
[ZHA 14]	Shared-memory supercomputer (24 cores)	OpenMP	~8

Table I.1. *Parallelization of DEM in different hardware platforms using various software programming environments*

Parallelization of a DEM code is a typical interdisciplinary task that requires not only deep understanding of the DEM but also some fundamental knowledge of computer science. Existing literature, for example those listed works in Table I.1, focuses only on some specific aspects of parallelization implementation of a DEM code. Details on the implementation are usually not well explained. Moreover, some parallel implementations of the DEM are too complex and betray the merit of the DEM being easily understood. In this book, a DEM code will be implemented using MATLAB as the programming environment. Then, the Parallel Computing Toolbox® of MATLAB is adopted to parallelize the DEM code to different platforms, that is multi-core PC, GPU computer and cluster.

This book comprises four chapters along with an introduction. Here, in the introduction, DEM and HPC are presented with a discussion on the definition of HPC and on the objectives and scope of the book. Chapter 1 focuses on the implementation of a serial DEM code. Details of mathematical equations and corresponding computational implementation are also covered in this chapter. Chapter 2 presents multi-core parallelization of the serial DEM code developed in Chapter 1. GPU parallelization of the DEM code is described in Chapter 3. Finally, Chapter 4 introduces using DICE2D in a middle-sized cluster.

I.3. Conclusion

The DEM is becoming a popular research tool in many fields. Parallelization is an essential step to overcome its computational limitation and fully use the power of modern computers. Until now, many parallel DEM codes have been developed for various hardware platforms using different parallel software programming environments. Nevertheless, implementation details are usually not well described. This book aims to provide the implementation details of a serial DEM code and its parallelization to modern parallel computers

1

Serial Implementation

In this chapter, implementation details of a serial DEM code, DICE2D, are described. The target is to provide a programming environment for further parallel implementation of the DEM. Full aspects of DEM coding are covered, such as system design, data structure design, flowchart design, algorithm design and implementation. A number of benchmark examples are designed for verification and debugging purposes.

1.1. System design

Unlike the modern commercial DEM codes, which use the object-oriented design concept, DICE2D adopts the process-oriented design concept that is simple to understand and suitable for algorithm research. To release the workload of graphical user interface (GUI) design, MATLAB® was selected as the programming environment. It provides a number of built-in functions to display the particles and computational results. Moreover, the high interactive feature of MATLAB also makes it a good choice for algorithm development and the study of the DEM. The DEM aims to simulate the nature process; therefore, the code is named DICE2D. It is inspired from Einstein's comment on quantum mechanics: "...God does not throw dice". To run the code, the first step is to select a folder named DICE2D as the current work directory (Figure 1.1). Then, type "D2D(iEx)" into the command window to run the corresponding example.

Figure 1.2 shows the work flow of DICE2D. First, the user needs to input the example ID. Then, the preprocessor will prepare the corresponding model data. The numerical model will be further processed by the DEM solver to obtain the simulation results, which will be finally processed by the postprocessor. Unlike most commercial software, in DICE2D, users must build up their own pre- and postprocessors to run a new example. The most convenient way is to modify the existing pre- and postprocessor files. For example the user can modify the existing preprocessor file to create different material parameters, loading conditions and particle models. Details of the pre- and postprocessors can be found in the source codes provided along with this book. In this chapter, the data structures and algorithms of DICE2D are introduced. It is well known that the classical equation of process-oriented design is Programs = Data Structures + Algorithms. Moreover, understanding the data structures and algorithms of DICE2D will bring deep insight into the work principle of the DEM.

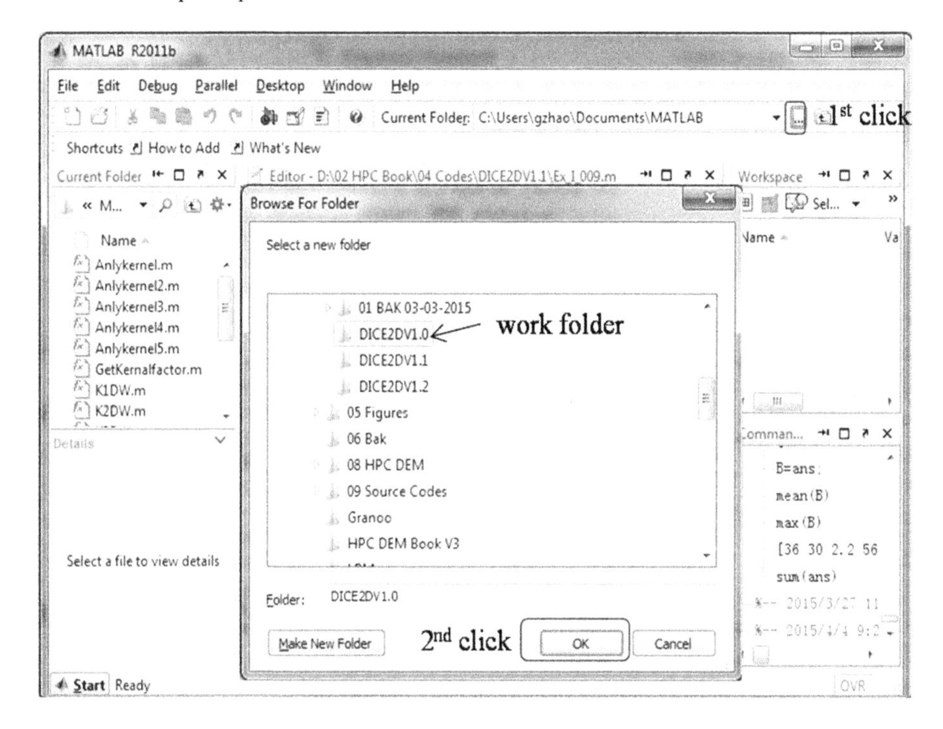

Figure 1.1. *DICE2D in MATLAB*

1.1.1. *Data structures*

Two essential entities of the DEM are particles and walls (Figure 1.3). Particles are usually used to represent the modeling target, such as a specimen and engineering structure; walls are adopted for applying boundary conditions. In DICE2D, the position, velocity, acceleration and forces of particles are defined as one-dimensional arrays. To consider additional degrees of freedom (DOFs), for example fluid flow or temperature, new arrays can be added. For walls, only position and velocity are defined, which can be used to apply velocity boundary conditions. For example if the velocities of a wall are set to zero, then the wall would be fixed during the computation. To apply a stress boundary condition, additional variables and treatments are needed. One example is presented in Chapter 3 to show the work principle of stress-controlled walls for triaxial compression test simulation. For convenience, material parameters of the particles and walls are defined separately (Figure 1.4). To link a specific particle to the corresponding material properties, a material ID of the particle or wall is adopted. For the particles and walls, the parameters of the classical Mohr–Coulomb model are defined. In addition, the viscous and stiffness parameters are defined on the particles; however, neither is defined for the walls because they are assumed to be rigid.

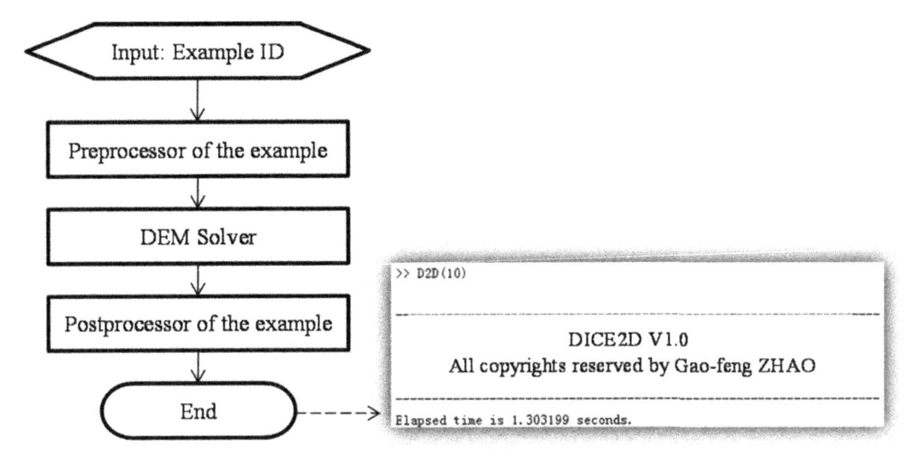

Figure 1.2. *Flowchart of DICE2D*

In many DEM simulations, walls are sufficient for applying types of boundary conditions. However, for some specific problems, precisely controlling the state of particles is required. To fulfill this requirement in DICE2D, particle boundary conditions are defined. The data structures are shown in Figure 1.5. These boundary conditions are defined at the particle level, which can control the force and displacement of a specific particle with a given ID. A special boundary condition called a fixed spring boundary condition is defined to simulate a fixed spring-like condition. The principle is to put a normal spring between the fixed point and the particle center. These particle boundary conditions, together with the walls, equip DICE2D with the ability to simulate problems that involve nonlinear deformation, progressive failure and granular flow. Table 1.1 shows a few variables defined for particles in DICE2D. Additional data structures of DICE2D are shown in Figure 1.6. These additional parameters include the threshold value for contact detection, total number of calculation cycles, measure point data, etc. These data are essential information for the preprocessor, solver and postprocessor. With these data structures, the skeleton of DICE2D is ready. The next step is to update these variables in each calculation cycle according to the DEM algorithms.

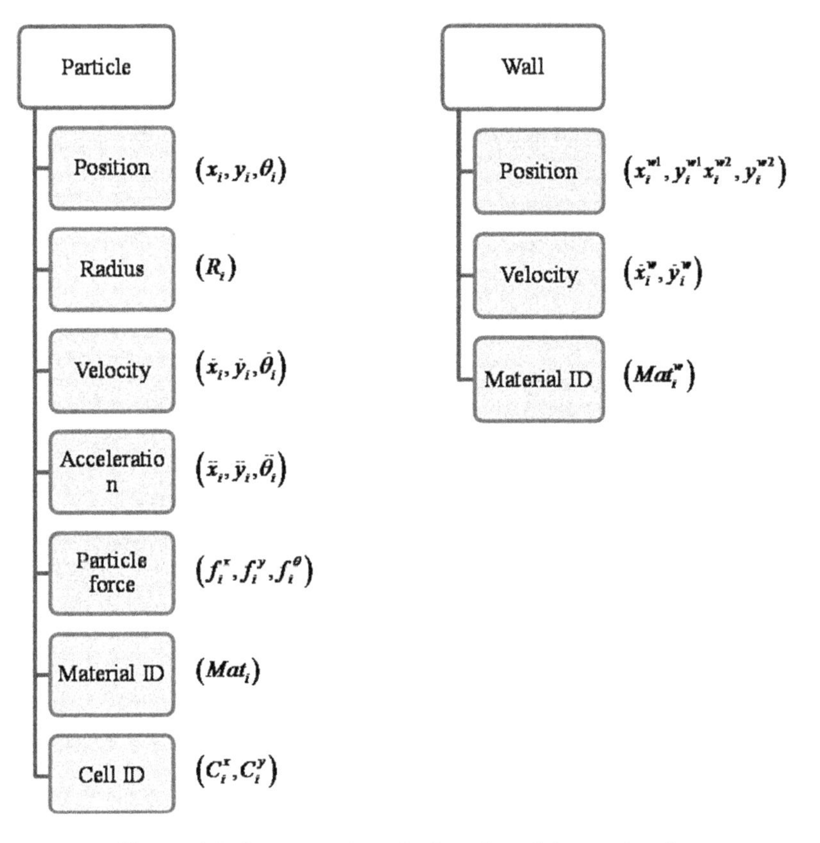

Figure 1.3. *Data structure design of particles and walls*

Variable	Description
NumP	Number of particles
X	Vector of particles' X coordinate (1 × NumP)
Y	Vector of particles' Y coordinate (1 × NumP)
R	Vector of particles' radius (1 × NumP)
T	Vector of particles' rotation (1 × NumP)
IDGX	Vector of particles' Grid ID in X direction
IDGY	Vector of particles' Grid ID in Y direction

Table 1.1. *Variables defined for particles in DICE2D*

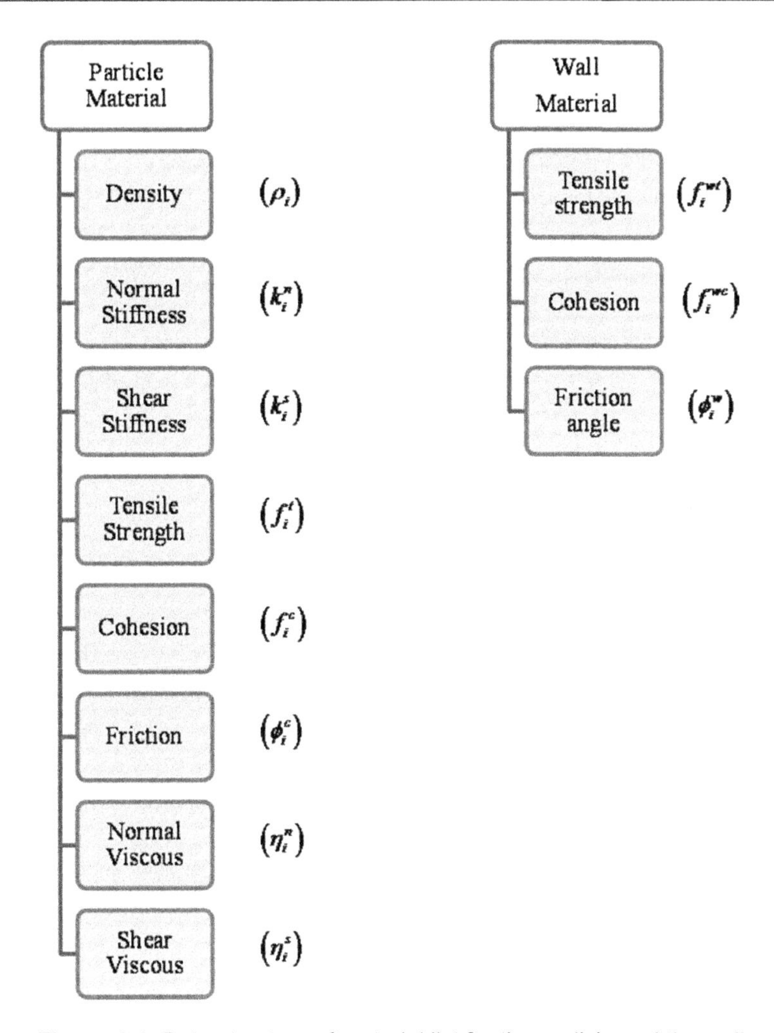

Figure 1.4. *Data structure of material list for the particle and the wall*

1.1.2. *Algorithms*

The data flow diagram of the DEM solver is shown in Figure 1.7. The data of particles and walls at time t are given first. Then, contact detection between particles and walls is conducted to obtain the contact pair lists for calculating the particle forces from the particle position and the constitutive models. The particle-to-particle (P2P) constitutive model is used to calculate the force–deformation relationship between particles, whereas the wall-to-particle (W2P) model is used to handle the interactions between the wall and the particles. When the particle forces are obtained, the state of particles at time $t+\Delta t$ can be updated according to Newton's second law of motion.

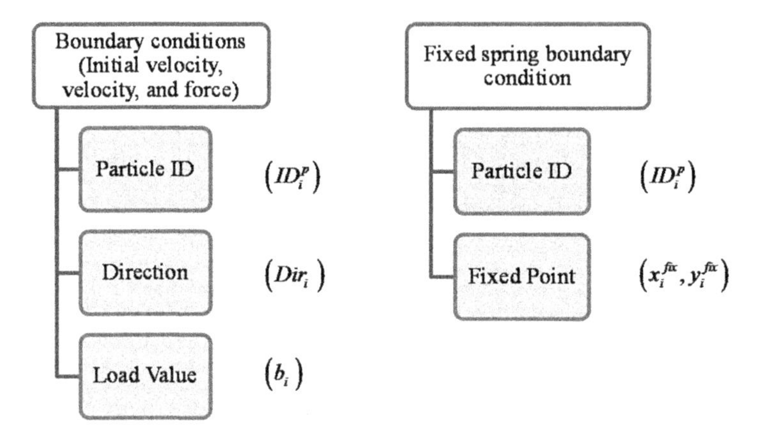

Figure 1.5. *Data structures of particle boundary conditions*

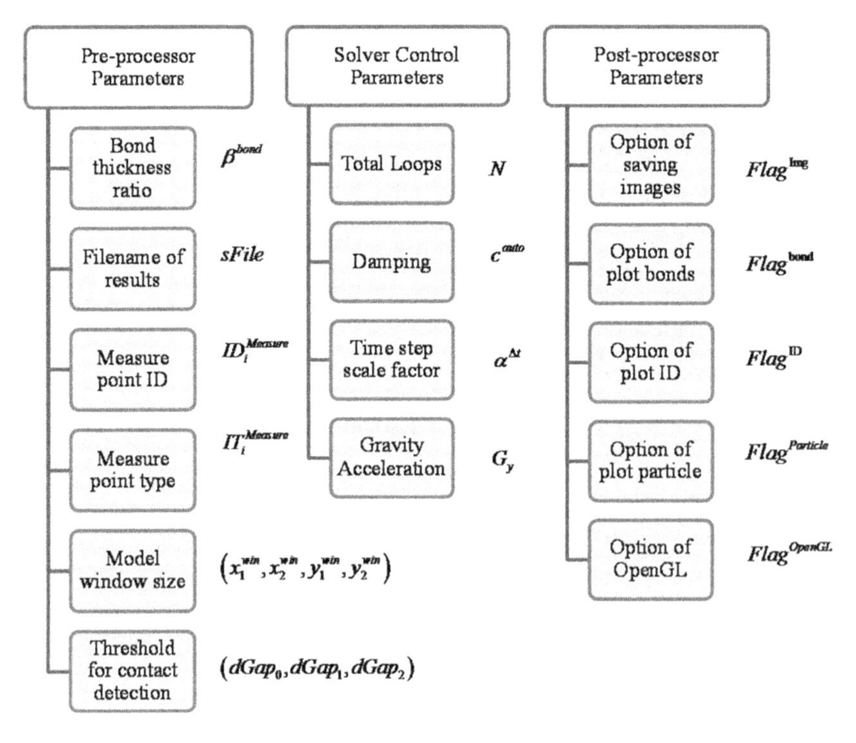

Figure 1.6. *Data structures for control parameters of preprocessor, solver and postprocessor*

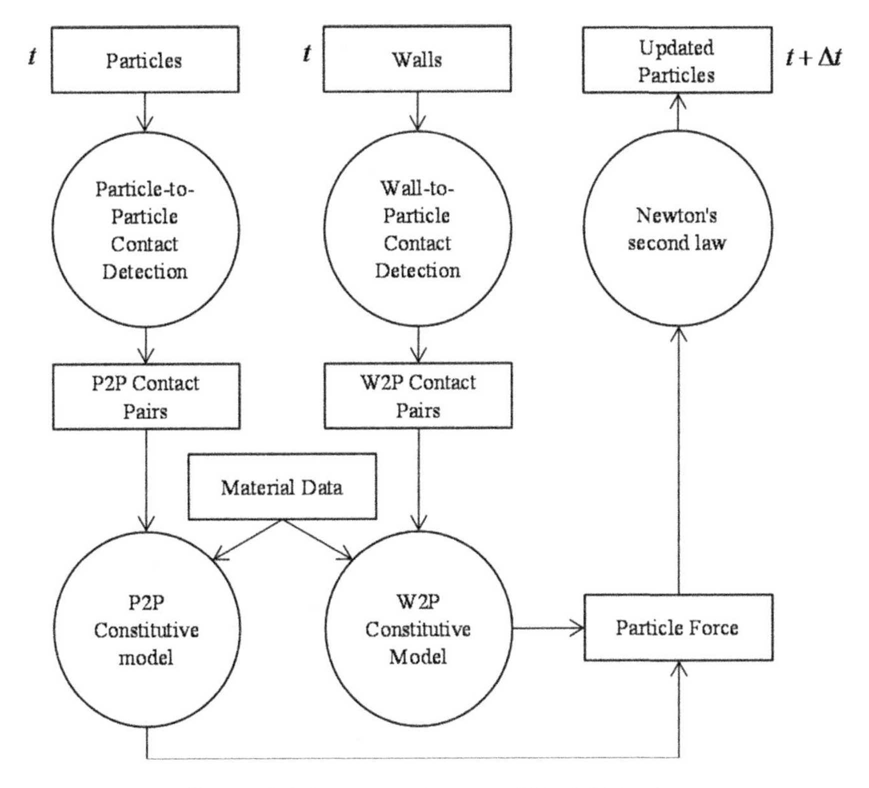

Figure 1.7. *Data flow diagram of the DEM solver*

In DICE2D, the mathematical equations of the velocity update of particles are as follows:

$$\dot{x}_i^{t+\Delta t/2} = \dot{x}_i^{t-\Delta t/2} + \frac{f_i^x}{m_i}\Delta t \qquad [1.1]$$

$$\dot{y}_i^{t+\Delta t/2} = \dot{y}_i^{t-\Delta t/2} + \frac{f_i^y}{m_i}\Delta t \qquad [1.2]$$

$$\dot{\theta}_i^{t+\Delta t/2} = \dot{\theta}_i^{t-\Delta t/2} + \frac{f_i^\theta}{I_i}\Delta t \qquad [1.3]$$

where $\dot{x}_i^{t+\Delta t/2}$ is the particle velocity in the *x*-direction at $t+\Delta t/2$; $\dot{x}_i^{t-\Delta t/2}$ is the particle velocity in the *y*-direction at $t-\Delta t/2$; f_i^x is the particle force in the *x*-direction at current time *t*; m_i is the mass of the particle and Δt is the time step, and other parameters in the *y*-direction and rotation direction are defined analogously.

The particle mass and moment of inertia can be obtained as follows:

$$M_i = \rho_i \pi R_i^2 \quad [1.4]$$

$$I_i = \frac{M_i R_i^2}{2} \quad [1.5]$$

where ρ_i is the density and R_i is the radius of the particle.

The updated positions of the particles are obtained as follows:

$$x_i^{t+\Delta t} = x_i^t + \dot{x}_i^{t+\Delta t/2} \Delta t \quad [1.6]$$

$$y_i^{t+\Delta t} = y_i^t + \dot{y}_i^{t+\Delta t/2} \Delta t \quad [1.7]$$

$$\theta_i^{t+\Delta t} = \theta_i^t + \dot{\theta}_i^{t+\Delta t/2} \Delta t \quad [1.8]$$

These equations constitute the motion update core of the DEM solver, which can easily be implemented. In Figure 1.7, the influence of the boundary condition was not considered. For particles with prescribed boundary conditions, the motion can be directly controlled by assigning the prescribed velocities or displacement. The implementation of the constitutive model with contact detection is complex. As mentioned in the Introduction, the constitutive model includes contact detection. To explain the implementation straightforwardly, contact detection is introduced first. In the following sections, the contact detection and the constitutive model developed in DICE2D are described.

1.2. Contact detection

1.2.1. *Simplified grid cell method*

Many algorithms have been developed for contact detection of the DEM; details on these methods can be found in the book by Munjiza [MUN 04]. In DICE2D, a simplified grid cell method is used. Rather than storing the particle ID in a grid cell, the grid cell numbers are stored in each particle. For a given particle, its cell numbers are assigned according to the following equations:

$$\mathrm{ID}_i^x = \left\lfloor \frac{x_i - \left(\min\left(x_i\right) - \max\left(R_i\right)\right)}{d\mathrm{GridSize}} \right\rfloor \qquad [1.9]$$

$$\mathrm{ID}_i^y = \left\lfloor \frac{y_i - \left(\min\left(y_i\right) - \max\left(R_i\right)\right)}{d\mathrm{GridSize}} \right\rfloor \qquad [1.10]$$

where $d\mathrm{GridSize}$ is the grid cell size, which can be estimated as $2\max(R_i)$. Between two particles, contact detection will first be checked if it satisfies the following conditions:

$$\left|\mathrm{ID}_i^x - \mathrm{ID}_j^x\right| < 2 \text{ and } \left|\mathrm{ID}_i^y - \mathrm{ID}_j^y\right| < 2 \qquad [1.11]$$

This prejudgment will eliminate unnecessary calculations and reduce the computational time (Figure 1.8), while the implementation and memory requirement are still simple.

In DICE2D, a buffer strategy is adopted to further reduce the computational time of contact detection. For each contact detection, potential contact pairs (particles with large threshold values) are detected and stored. This will result in the code not needing to perform contact detection in a few future steps. Contact detection will only be triggered when the maximum accumulated displacement of the particles is larger than the given threshold value. This will lead to some DEM simulations, for example continuum and fracturing with small deformation, only needing to perform contact detection at the beginning. Because these problems are the main concerns of rock mechanics, it was decided that DICE2D should take the simple grid cell method rather than these more complex counterparts. In the following sections, details of the contact detection of P2P and W2P are presented.

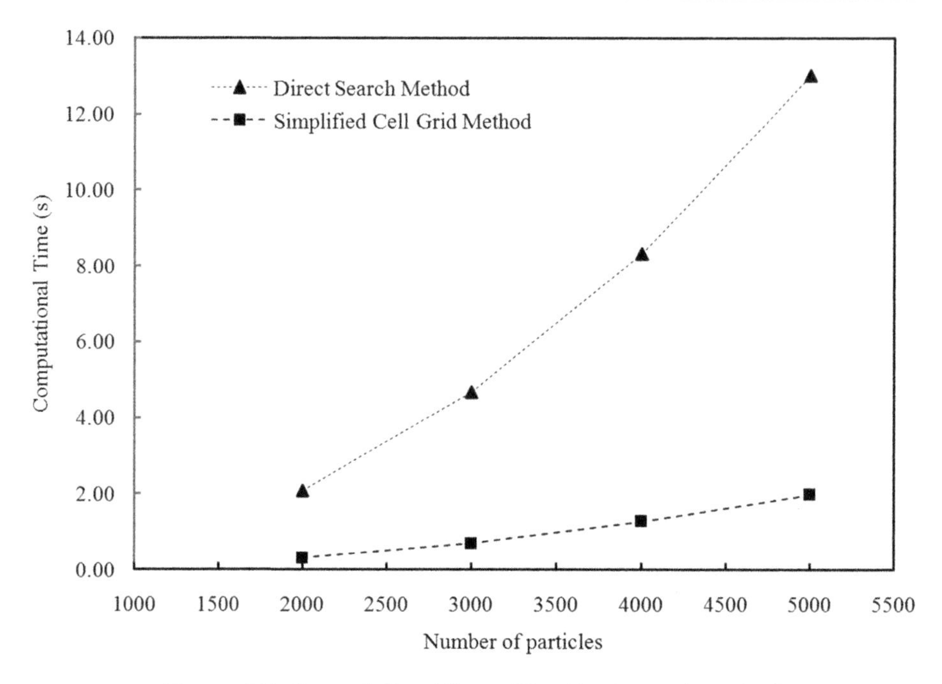

Figure 1.8. *Computational time of the direct search method and simplified grid cell method*

1.2.2. *Particle-to-particle contact*

The judgment of two particles in contact is expressed as follows:

$$\sqrt{(x_i - x_j)^2 + (y_i - y_j)^2} - R_i - R_j < d\text{Gap}_0 \quad [1.12]$$

where $d\text{Gap}_0$ is a threshold value to form a contact pair or potential contact pair.

In DICE2D, a global P2P contact list of the whole model is further divided into segments for the individual particle. For each particle, there is an array of fixed size to store its contact pairs (10 is the default size). Figure 1.9 shows contact detection for initial contact pairs, which includes two steps. The first step is contact detection with a small threshold value $(d\text{Gap}_1)$ to obtain the initial contact pair list. The second step is potential contact pair detection using a bigger threshold value $(d\text{Gap}_2)$. During the second contact detection step, a deleting operation is first performed to remove contact pairs whose gaps are larger than the threshold value $(dGap_2)$. Following this, a search of existing contact pairs is performed. Only new pairs will be added to the contact list. More specifically, the potential contact pair will first be checked to see whether it is already listed. In the meantime, the algorithm will find one empty space on the list. If the contact pair is new and there is an empty space, then it will be allocated to the empty space.

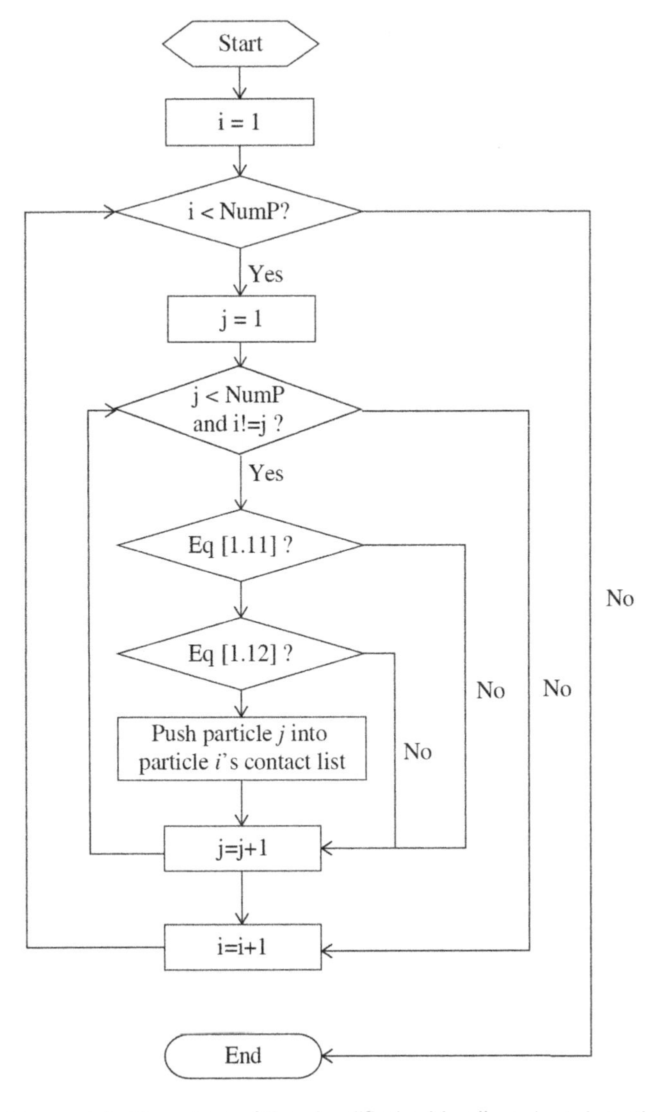

Figure 1.9. *Flowchart of the simplified grid cell contact detection*

Contact detection during the calculation involves only the second step. It will be triggered when the accumulative displacement of a particle is larger than the given trigger threshold value ($0.3d\text{Gap}_2$ in DICE2D). When contact detection is performed, the accumulative displacement will be reset to zero. This conditional contact detection will largely reduce the computational time for the contact detection of cohesive granular material.

1.2.3. *Wall-to-particle contact*

The contact between particle and wall is more complex. There are two possible situations: one is the particle-to-wall case and the other is the particle-to-wall-end case. Figure 1.10 shows the contact condition of the first case. Mathematically, it can be written as follows:

$$\begin{cases} 0 < \overrightarrow{BC} \cdot \overrightarrow{BA} < \left|\overrightarrow{BC}\right| \\ \left|\overrightarrow{DA}\right| - R < d\text{Gap} \end{cases} \qquad [1.13]$$

where $\overrightarrow{BC}$ is the vector of the wall, $\overrightarrow{BA}$ is the vector from one end of the wall to the particle, $\left|\overrightarrow{BC}\right|$ is the length of the wall, $\left|\overrightarrow{DA}\right|$ is the distance from the particle to the wall, R is the radius of the particle and dGap is the threshold value of contact pair formation. When the particle and the wall are in contact, the normal direction is calculated as follows:

$$n^c = \frac{\overrightarrow{DA}}{\left|\overrightarrow{DA}\right|} \qquad [1.14]$$

where $\overrightarrow{DA} = \overrightarrow{BA} - \overrightarrow{BC}\left(\overrightarrow{BC} \cdot \overrightarrow{BA} / \left|\overrightarrow{BC}\right|\right)$.

The particle-to-wall-end contact happens when:

$$\begin{cases} \overrightarrow{BC} \cdot \overrightarrow{BA} \le 0 \ \text{ or } \ \overrightarrow{BC} \cdot \overrightarrow{BA} \ge \left|\overrightarrow{BC}\right| \\ \left|\overrightarrow{BA}\right| - R < d\text{Gap} \ \text{ or } \ \left|\overrightarrow{CA}\right| - R < d\text{Gap} \end{cases} \qquad [1.15]$$

In this case, contact normal direction is calculated as follows:

$$n^c = \begin{cases} \dfrac{\overrightarrow{BA}}{\left|\overrightarrow{BA}\right|} & \left|\overrightarrow{BA}\right| - R < d\text{Gap} \\ \dfrac{\overrightarrow{CA}}{\left|\overrightarrow{CA}\right|} & \left|\overrightarrow{CA}\right| - R < d\text{Gap} \end{cases} \qquad [1.16]$$

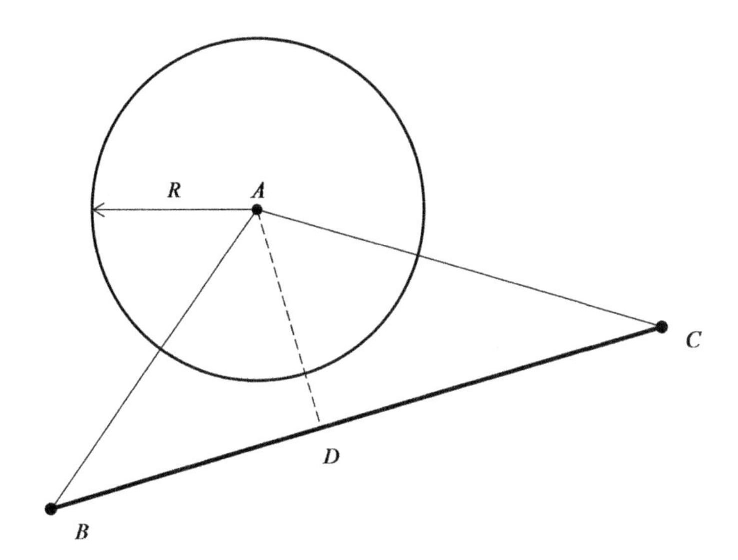

Figure 1.10. *Contact detection between wall and particle*

Rather than N^2 complexity of the P2P contact detection, because the number of walls in a simulation is usually much less than the number of particles, the complexity of W2P contact detection is an N complexity algorithm. Therefore, the direct search method is used for W2P contact detection in DICE2D. This algorithm also has two steps: the first step is to form the initial contact list, and the second step is to form potential contacts. During calculation, only the second step is used to update the W2P contact list. The trigger condition of W2P contact detection is the same as that of P2P contact detection. It should be mentioned that contact detection actually provides candidates only for interaction calculation. To obtain the actual particle forces, constitutive models are required for both P2P and W2P contact pairs.

1.3. Constitutive model

The constitutive model is the core of the DEM, providing a relationship between deformation of the particle contact and the particle contact forces. The ability of a DEM code is mainly determined by its constitutive model. For example the classical DEM with the Mohr–Coulomb model can be used to simulate granular materials. To model the fracturing of brittle rock, an enriched bond constitutive model is required. In this chapter, the constitutive model of DICE2D is represented in an integrated way. It is a bond model when the material parameters (tensile and cohesion strength) are non-zero values; otherwise, it is a classical Mohr–Coulomb contact model.

1.3.1. *Particle-to-particle constitutive model*

Figure 1.11 shows the mechanical components of the P2P contact. Compared with the classical model of the DEM, the normal spring is further divided into two subsprings that are distributed with a distance d^b. This is a simplified version of the bond model proposed by Potyondy and Cundall [POT 04], which used the rolling model concept of Jiang *et al.* [JIA 05]. Instead of continuous integration along the interface, only two subnormal springs are adopted here. When d^b is non-zero, the model is a bond model that can transmit a moment between two particles; otherwise, it is the classical Mohr–Coulomb contact model in the DEM. In addition to springs, two dashpots are imposed between two particles along the normal and shear directions. For the shear direction, because a separate distribution of subsprings would have no influence on the final mechanical response, only a single spring is imposed. To obtain the particle interaction force, an incremental form is adopted to link the force increment with the deformation increment (velocity).

For a P2P contact, the relative angular velocity between two particles is as follows:

$$\dot{\theta}_{ij} = \dot{\theta}_j - \dot{\theta}_i \qquad [1.17]$$

The corresponding forces induced at the two subnormal springs are calculated as follows:

$$\Delta F_n^{\Theta} = \pm \frac{1}{4} \dot{\theta}_{ij} \Delta t d^b k_n \qquad [1.18]$$

where k_n is the normal stiffness.

The corresponding moment is calculated as follows:

$$\Delta M_{ij}^{\Theta} = \frac{1}{4}\dot{\theta}_{ij}\Delta t d^{b2} k_n \qquad [1.19]$$

Between two particles, the two subsprings are in a group; therefore, the angular velocity–induced global normal force between two particles is canceled to zero.

The normal velocity of the P2P contact is calculated as follows:

$$v_n = (\dot{x}_j - \dot{x}_i)n_x + (\dot{y}_j - \dot{y}_i)n_y \qquad [1.20]$$

where

$$n_x = \frac{x_j - x_i}{\sqrt{(x_j - x_i)(x_j - x_i) + (y_j - x_i)(y_j - y_i)}}$$
$$n_y = \frac{y_j - y_i}{\sqrt{(x_j - x_i)(x_j - x_i) + (y_j - x_i)(y_j - y_i)}} \qquad [1.21]$$

The normal force increment between two particles is obtained as follows:

$$\Delta F_n = k_n v_n \Delta t \qquad [1.22]$$

Finally, together with the normal and angular velocities, for each subspring, its normal force can be written as follows:

$$\Delta \tilde{F}_n^{\Theta} = \frac{1}{2}k_n v_n \Delta t \pm \frac{1}{4}\dot{\theta}_{ij}\Delta t d^b k_n \qquad [1.23]$$

For the shear direction of the P2P contact, its relative velocity is calculated as follows:

$$v_s = -(\dot{x}_j - \dot{x}_i)n_y + (\dot{y}_j - \dot{y}_i)n_x - \dot{\theta}_i R_i - \dot{\theta}_j R_j \quad [1.24]$$

The induced shear force can be given as follows:

$$\Delta F_s = k_s v_s \Delta t \quad [1.25]$$

where k_s is the shear stiffness.

The shear force distributed to the shear springs linked to each subnormal spring is given as follows:

$$\Delta \tilde{F}_s^{\Theta} = \frac{1}{2} k_s v_s \Delta t \quad [1.26]$$

For each subspring pair, the Mohr–Coulomb model with a tension cut is adopted (Figure 1.12). The total normal and shear forces of a P2P contact pair are integrated along time as follows:

$$\tilde{F}_n^{\Theta,t} = \tilde{F}_n^{\Theta,t-\Delta t} + \Delta \tilde{F}_n^{\Theta} \quad [1.27]$$

$$\tilde{F}_s^{\Theta,t} = \tilde{F}_s^{\Theta,t-\Delta t} + \Delta \tilde{F}_s^{\Theta} \quad [1.28]$$

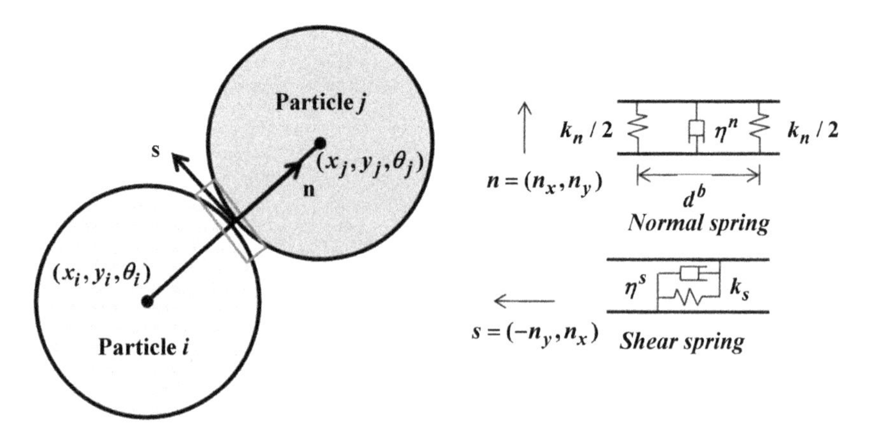

Figure 1.11. *Contact detection between the wall and particle*

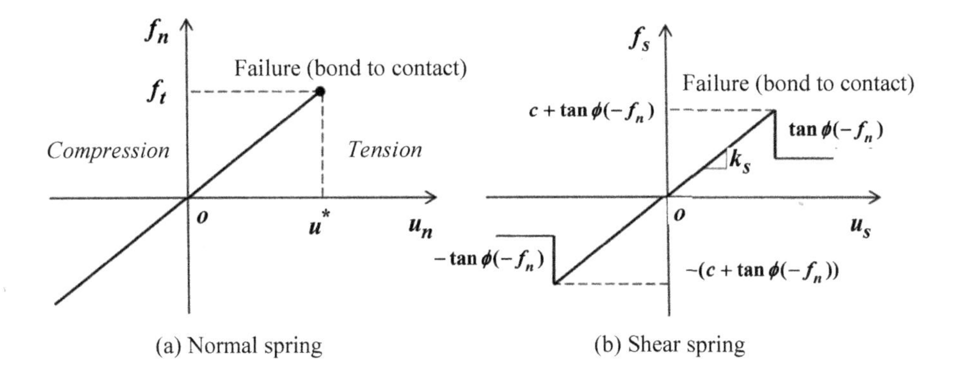

Figure 1.12. *Constitutive model of the P2P contact*

The earlier-mentioned equations are correct in an elastic range. The contact bond will be broken when the following conditions are satisfied:

$$\tilde{F}_n^{\Theta,t} \geq \frac{1}{2} F_t \qquad [1.29]$$

or

$$\left|\tilde{F}_s^{\Theta,t}\right| \geq \tilde{F}_n^{\Theta,t} \tan\phi + \frac{1}{2} F_c \qquad [1.30]$$

where F_t is the tensile strength and F_c is the cohesion strength of the contact.

The P2P contact pair will turn into a Mohr–Coulomb contact when the aforementioned condition is satisfied at least for one subspring. When failure occurs, the two subnormal springs will be merged into a single normal spring. Then, the moment accumulated between the particles will be released to zero. The cohesion and tension strength of the contact pair will be set to zero. Finally, the normal force and the shear force between the newly released contacts will be recalculated using the Mohr–Coulomb model (Figure 1.13).

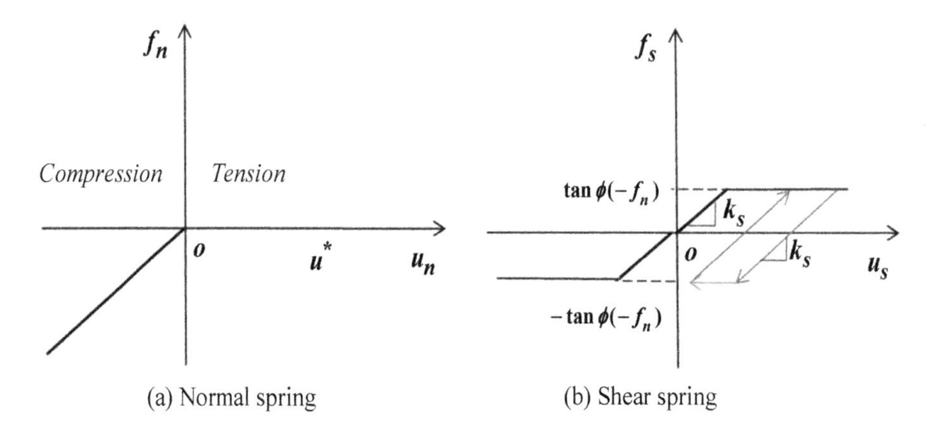

(a) Normal spring (b) Shear spring

Figure 1.13. *Constitutive model of the normal Mohr–Coulomb contact*

In DICE2D, interaction forces between two particles include one normal force, one shear force and one moment $(\tilde{F}_n^t, \tilde{F}_s^t, M_s^{\Theta,t})$. When contact pair failures occur, the last term is set to zero and would not involve further calculation. The stiffness parameters of the contact are not changed. It should be mentioned that only the initial contact detection procedure can generate bonded P2P contact pairs. Contact detection during calculation can produce only Mohr–Coulomb contact pairs. When the material parameters (tensile and cohesion) are zero, the initial P2P contact pairs are also treated as Mohr–Coulomb contact pairs. Using this P2P constitutive model, DICE2D can model both cohesive and non-cohesive granular materials.

In DICE2D, the dashpot's viscous forces are calculated as follows:

$$F_n^{vs} = \eta_n k_n v_n \quad [1.31]$$

$$F_s^{vs} = \eta_s k_s v_s \quad [1.32]$$

where η_n and η_s are dimensionless normal and shear viscous parameters (unit s), respectively. Viscous forces are only considered for P2P pairs that are in bond state or contact state. When the contact pair is separated, the viscous forces are set to zero.

Until now, all contact forces were defined according to local coordinates. The particle force and the moment contributed from each P2P contact pair under the whole coordinate system are given as follows:

$$F_i^x = \left(F_n + F_n^{vs}\right)n_x^c + \left(F_s + F_s^{vs}\right)n_y^c \qquad [1.33]$$

$$F_i^y = -\left(F_n + F_n^{vs}\right)n_y^c + \left(F_s + F_s^{vs}\right)n_x^c \qquad [1.34]$$

$$M_i^y = \left(F_s + F_s^{vs}\right)R - M^{\Theta} \qquad [1.35]$$

The total particle forces can be obtained from a sum operation of all forces contributed by the P2P contacts. For particles in contact with walls, the contribution from the W2P contacts must be considered. In the following section, the W2P constitutive model is presented.

1.3.2. *Wall-to-particle constitutive model*

In W2P contact, there are two springs: one is a normal spring and another is a shear spring (Figure 1.14). The contact pair is first assigned as Mohr–Coulomb with a tension cut (see Figure 1.12). When the contact pair is broken, it will then follow the Mohr–Coulomb model, as shown in Figure 1.13. Because the moment between particles and the wall can be automatically considered from multiple contacts (see Figure 1.14), the W2P contact does not consider the bond thickness.

According to the contact detection of the W2P contact, there are three possible contact points, that is *B*, *D* or *C* (see Figure 1.10). In the W2P constitutive model, the first step is to determine the contact point (x_i^{wc}, y_i^{wc}). For actual implementation, the following procedure is used: if $\left|\overrightarrow{BA}\right| - R < d\text{Gap}$, then the contact point is *B* and exit; if $\left|\overrightarrow{CA}\right| - R < d\text{Gap}$, then the contact point is *C* and exit; if equation [1.13] is satisfied, then the contact point is *D*.

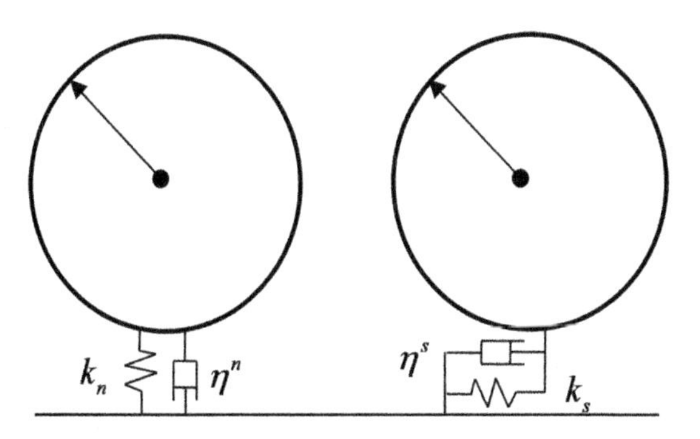

Figure 1.14. *Components of the W2P contact*

The contact direction of the W2P pair is calculated as follows:

$$n_x^{cw} = \frac{x_j - x_i^{wc}}{\sqrt{(x_j - x_i^{wc})(x_j - x_i^{wc}) + (y_j - y_i^{wc})(y_j - y_i^{wc})}} \quad [1.36]$$

$$n_y^{cw} = \frac{y_j - y_i^{wc}}{\sqrt{(x_j - x_i^{wc})(x_j - x_i^{wc}) + (y_j - y_i^{wc})(y_j - y_i^{wc})}} \quad [1.37]$$

The normal deformation of the W2P contact is calculated as follows:

$$\delta_n = \sqrt{(x_j - x_i^{wc})(x_j - x_i^{wc}) + (y_j - y_i^{wc})(y_j - y_i^{wc})} - R_j \quad [1.38]$$

If the pair is in contact state ($\delta_n < 0$), then the normal force is given as follows:

$$F_n^{\mathrm{P2W}} = \delta_n k_n^{\mathrm{P2W}} \quad [1.39]$$

where k_n^{P2W} is the normal stiffness of the contact pair, which is twice the normal stiffness of the linked particle.

The shear force is calculated using an incremental method. First, the relative velocity of the W2P contact along the shear direction is calculated as follows:

$$v_s^{\mathrm{P2W}} = -(v_j^x - v_i^{wx}) n_y^{cw} + (v_j^y - v_i^{wy}) n_x^{cw} - \dot{\theta}_j R_j \quad [1.40]$$

$$v_n^{\mathrm{P2W}} = (v_j^x - v_i^{wx}) n_x^{cw} + (v_j^y - v_i^{wy}) n_y^{cw} \quad [1.41]$$

where v_i^{wx} and v_i^{wy} are velocities of the wall. The incremental shear force is obtained as follows:

$$\Delta F_s^{\mathrm{P2W}} = k_s^{\mathrm{P2W}} \Delta t k_s^{\mathrm{P2W}} \quad [1.42]$$

where k_s^{P2W} is the shear stiffness of the contact. Similar to the P2P contact, the total shear force of the W2P contact needs to be integrated along time and modified according to the prescribed constitutive model. If the W2P contact fails, both the tensile and the cohesion strengths of the contact would be set to zero. When a W2P contact is in bond state (tensile and cohesion strengths are not zero), both the normal and the shear forces are calculated incrementally. However, when it is broken, the normal force will be calculated using equation [1.30], whereas the shear force will be calculated incrementally as shown in equation [1.42].

The viscosity between the particle and the wall is also considered using the following equations:

$$F_n^{wvs} = \eta_n^{\mathrm{P2W}} k_n v_n \qquad [1.43]$$

$$F_s^{wvs} = \eta_s^{\mathrm{P2W}} k_s v_s \qquad [1.44]$$

where η_n^{P2W} and η_s^{P2W} are the viscous parameters in the normal and the shear directions of the P2W contact, respectively (the same viscous coefficients as the linked particle).

The viscous forces will only be active when the W2P contact pair is in bond or contact state. The particle force contributed from a W2P contact under the whole coordinate system is written as follows:

$$F_i^x = \left(F_n^{\mathrm{P2W}} + F_n^{wvs}\right) n_x^{cw} + \left(F_s^{\mathrm{P2W}} + F_s^{wvs}\right) n_y^{cw} \qquad [1.45]$$

$$F_i^y = -\left(F_n^{\mathrm{P2W}} + F_n^{wvs}\right) n_y^{cw} + \left(F_s^{\mathrm{P2W}} + F_s^{wvs}\right) n_x^{cw} \qquad [1.46]$$

$$M_i^y = \left(F_s^{\mathrm{P2W}} + F_s^{wvs}\right) R \qquad [1.47]$$

The forces from the W2P contact pair list can be further added to the particles using a sum operation. Until now, essential theoretical parts of the DEM code, for example main framework, motion equations, contact detection and constitutive model, have been presented. However, for a complete DEM code, some additional functions are required, for example a damping scheme for solving the static problem, selection of a time step and energy calculation for postprocessing. These issues are covered in the following section.

1.4. Time step, damping and energy calculation

The DEM is an explicit numerical method. To keep the computation stable, a small time step is required. In DICE2D, the time step is estimated as follows:

$$\Delta t = \alpha \min\left(2\sqrt{\frac{M_i}{\max(k_i^n, k_i^s)}} \right) \quad [1.48]$$

where α is a reduction coefficient (0.1 as default) to ensure numerical stability. Equation [1.48] is from the requirement that the time step in the explicit method must be less than the time needed for elastic wave propagation through the smallest element of the model. The reduction parameter α can be used to adjust the time step in a dimensionless way, which is straightforward in the study of the influence of the time step on DEM simulation.

For quasi-static problems, damping is needed in the DEM to achieve the equilibrium condition. The adaptive damping scheme developed by Cundall for the DEM was adopted and can be written as follows:

$$F_x = F_x - C^{\text{auto}} \operatorname{sign}(\dot{x}) |F_x| \quad [1.49]$$

$$F_y = F_y - C^{\text{auto}} \operatorname{sign}(\dot{y}) |F_y| \quad [1.50]$$

$$M_\theta = M_\theta - C^{\text{auto}} \operatorname{sign}(\dot{\theta}) |M_\theta| \quad [1.51]$$

where C^{auto} is the local damping coefficient, with a suggested value of 0.8 for the static problem. This results in an overdamped system. In some cases, a smaller value might produce better results. It should be mentioned that the latest version of Particle Flow Code (PFC) suggests 0.6 as the default value [ITA 08].

For all DEM simulation, it is necessary to check the energy balance. Moreover, energy analysis is also an important postprocess of numerical simulation results. In DICE2D, three energies of the computational model are recorded during calculation, they are the kinematic energy, the strain energy and the gravity positional energy:

$$\Pi^{\text{kinematic}} = \sum_{\text{particles}} \frac{1}{2} M_i \left(v_i^x v_i^x + v_i^y v_i^y + v_i^y v_i^y \right) + \frac{1}{2} I_i \dot{\theta}_i \dot{\theta}_i \quad [1.52]$$

$$\begin{aligned} \Pi^{\text{strain}} = & \sum_{\text{P2P pairs}} \frac{1}{4} \frac{F_n^A}{k_n} F_n + \frac{1}{4} \frac{F_n^B}{k_n} F_n + \frac{1}{2} \frac{F_s}{k_s} F_s \\ & + \sum_{\text{P2W pairs}} \frac{1}{2} \frac{F_n^w}{k_n^w} F_n^w + \frac{1}{2} \frac{F_s^w}{k_s^w} F_s^w \end{aligned} \quad [1.53]$$

$$\Pi^{\text{gravity}} = \sum_{\text{particles}} M_i y_i g \qquad [1.54]$$

where F_n^A and F_n^A are the forces of two subnormal springs.

1.5. Benchmark examples

The DEM is relatively new, and some researchers regard it as a "not yet proven" numerical method (e.g. [ZHA 08]). Therefore, fundamental verifications of a DEM code are necessary before conducting any engineering applications. Moreover, for a new DEM code, debugging is always unavoidable. This can be realized only from running a number of specially designed numerical examples. In this section, a number of benchmark examples for DICE2D are presented, which are so called because the main purpose is to check the fundamental algorithmic implementation of DICE2D.

1.5.1. *Falling ball under gravity*

In this example, the problem of a ball falling down under gravity is simulated using DICE2D. Assume that the original position of the ball is zero and let it fall under gravity. Its falling distance can be calculated as follows:

$$y(t) = -\frac{1}{2} g t^2 \qquad [1.55]$$

where g is the gravitational acceleration and t is the falling time. This example is free from contact detection and the contact constitutive law. It involves only gravity and the calculation cycle described by equations [1.1–1.8]. The simulation results are controlled by Newton's second law of motion, which must be implemented correctly in the DEM code. Therefore, the example is the best candidate to check the implementation of the calculation core of a DEM code. It is suggested that such an example should be tested for any newly developed DEM code.

Number of particles	1	Shear stiffness (N/m)	1e9
Mean particle size (m)	10	Time step reduction factor	0.2
Density (kg/m^3)	1	Total steps	20,000
Normal stiffness (N/m)	1e9	Gravitational acceleration (m/s^2)	10

Table 1.2. *Model parameters of the falling ball problem*

The computational model has only one particle. The main model parameters used in the simulation are listed in Table 1.2.

Figure 1.15 shows the simulation results using DICE2D. As expected, the ball falls down under the effect of gravity. From this figure, a general conclusion on the correctness of the implementation of the calculation core and postprocessing in DICE2D can be achieved. If a DEM code cannot give the same results, then debugging should focus on the coding of equations [1.1–1.8]. Moreover, the postprocessing code should also be checked.

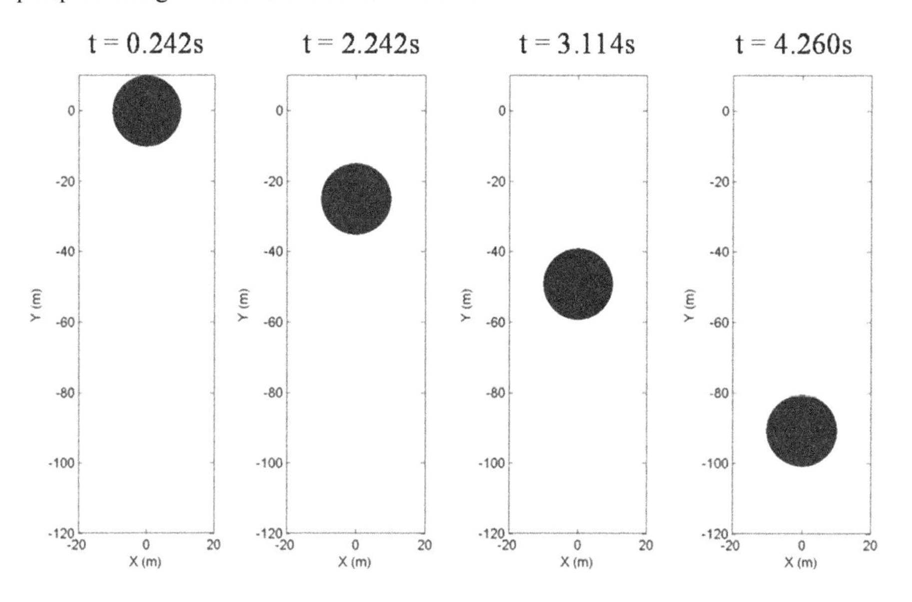

Figure 1.15. *Falling ball under gravity predicted using DICE2D*

Directly checking the final graphic outputs is a quick debugging method that is usually used during the coding period. To have a more quantitative comparison, the analytical results calculated using equation [1.55] are plotted together with the DEM prediction (Figure 1.16(a)). The close fit between the DEM and analytical solution can imply that equations [1.1–1.8] are precisely coded. Moreover, the energy analysis of the problem is shown in Figure 1.16(b). The energy calculation of the DEM can be determined from the figure. The energy equilibrium can ensure that the simulation is physically correct. From the figure, it is found that the strain energy of the system is zero from begin to end. This is because there is only one particle and there are no springs available to store the strain energy. Regarding the kinematic energy and gravity potential energy, the gravity potential energy decreases during the fall and changes into kinematic energy. As seen in Figure 1.16(b), an energy balance exists between these two energies.

For computer implementation, only one benchmark example can be used to check the data flow under specific conditions. For example the ball falling problem can check only the correctness of implementation of the particle movement in the vertical direction. In the following section, a more complex problem is used to test the implementation of DICE2D in both the vertical and the horizontal directions.

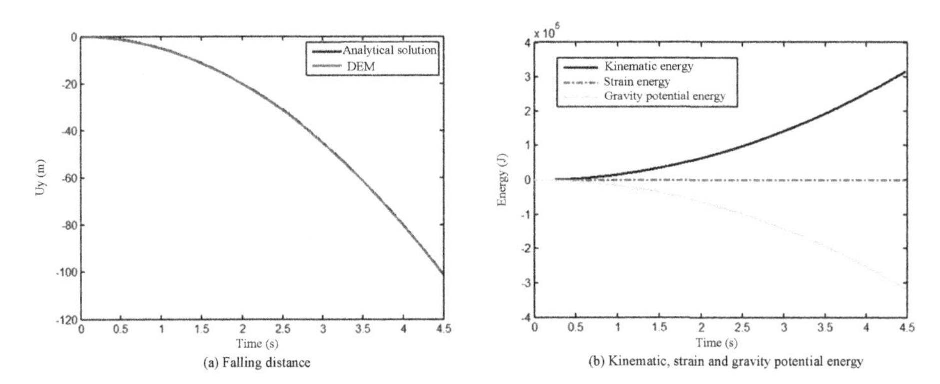

Figure 1.16. *Falling distance and energy analysis of the falling ball problem using DICE2D*

1.5.2. *Pendulum problem*

To check the implementation of DICE2D in terms of movement in both the vertical and the horizontal directions, the pendulum problem is selected as the second benchmark example. As shown in Figure 1.17, a ball is fixed with an inclined bar, which will move in both the vertical and the horizontal directions under the force of gravity. Its differential equation is written as follows:

$$\theta = -\frac{g}{L\sin\theta} \qquad [1.56]$$

where θ is the angle between the bar and vertical line, g is the gravitational acceleration and L is the length of the bar. Equation [1.56] can be solved using the central difference method or any other numerical integration method. The position of the ball in the y-direction, $L\cos(\theta)$, is adopted as the target variable to be compared with the DEM simulation.

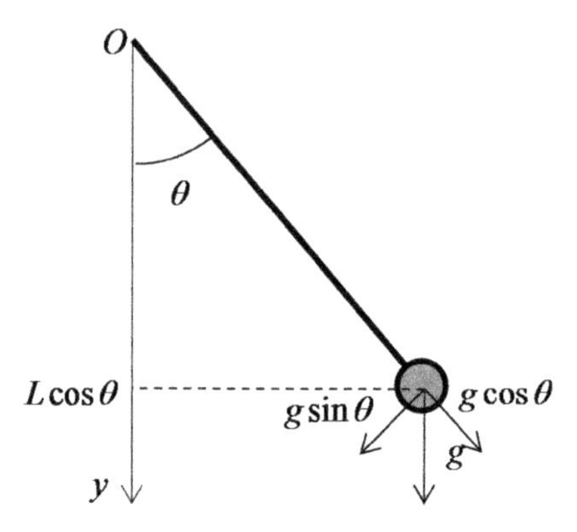

Figure 1.17. *The pendulum problem*

In this computational model, there is one particle with a coordinate of (100,0). The model parameters of the problem are shown in Table 1.3. Unlike the falling ball problem, a fixed spring boundary condition is applied to the particle, which is described by three numbers as [1 0 0]. The first number is the particle ID, the second one is the x-coordinate of the fixed point and the third number is the y-coordinate of the fixed point.

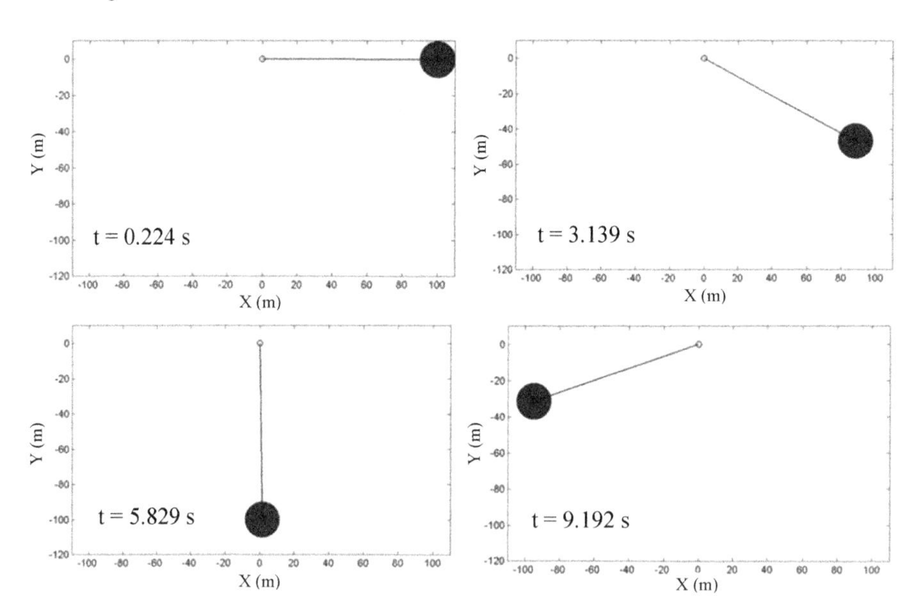

Figure 1.18. *Movement of the pendulum problem predicted using DICE2D*

Figure 1.18 shows the simulation results of DICE2D. The movement of the ball is well captured. It should be mentioned that the system is rotated by more than 180° during the calculation. This shows that the DEM is able to solve the large rotation problem directly. Unlike the large deformation finite element method (FEM) [HUG 80], no special treatment is required in the DEM for problems involving large rigid body rotation.

Number of particles	1	Shear stiffness (N/m)	1e9
Mean particle size (m)	10	Time step reduction factor	0.2
Density (kg/m^3)	1	Total steps	80,000
Normal stiffness (N/m)	1e9	Gravitational acceleration (m/s^2)	10

Table 1.3. *Model parameters of the pendulum problem*

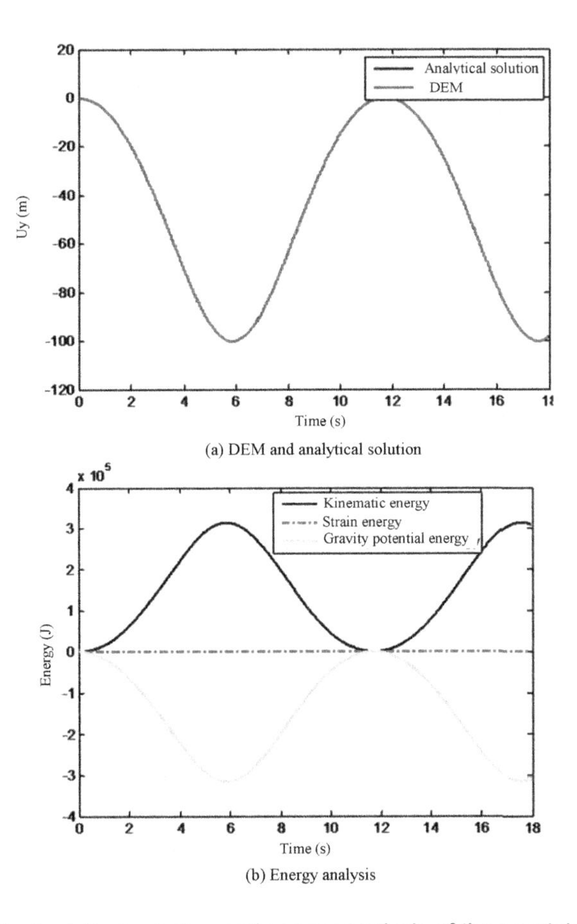

Figure 1.19. *Analytical solution and energy analysis of the pendulum problem*

Again, for a quantitative comparison, the analytical solution of equation [1.56] and the DEM prediction are plotted together in Figure 1.19(a). An exact fit is observed. This shows that the DEM can precisely solve this large rotation problem. Moreover, the energy analysis is shown in Figure 1.19(b). The kinetic and gravity potential energies vary in a cosine function of time, whereas the total energy is in equilibrium, that is the change in kinetic energy equals that of the gravity potential energy. From this example, the implementation of the calculation core in DICE2D is fully checked. From now on, during numerical simulation, if errors occur, debugging can focus on other parts. This example also tested the implementation of the fixed spring boundary condition. However, in both the aforementioned examples, only one particle is involved; the fundamental implementation of contact is still not considered. In the following section, the implementation of contact will be addressed step by step.

1.5.3. *Elastic deformation of normal spring under tension*

The calculation core and fixed spring boundary condition of DICE2D are checked in the aforementioned two examples. The targets of this example are to check the implementation of a normal spring under the viscous–elastic condition and force boundary condition. A simple tension test of a two-particle system is selected as the benchmark example. The computational model is shown in Figure 1.20. There are two particles; a contact pair will be formed during the calculation. The bottom particle (particle ID = 1) is fixed through a particle fixed boundary condition, described by three numbers as [1 2 0]. The first variable refers to the particle ID, the second to the direction (1 refers to x and 2 is y) and the third to the applied velocity. A tension force is applied to the top particle (particle ID = 2) using a particle force boundary condition, represented as [2 2 1e6]. The first two parameters represent the particle ID and direction, respectively. The last one is the applied force. The model parameters of the problem are shown in Table 1.4.

Number of particles	2	Normal viscous coefficient (s)	1e − 4
Mean particle size (m)	10	Time step reduction factor	0.2
Density (kg/m^3)	1	Total steps	2,000
Normal stiffness (N/m)	1e9	Gravitational acceleration (m/s^2)	0
Shear stiffness (N/m)	1e9		

Table 1.4. *Model parameters of the tension test*

The analytical solution of elastic deformation is as follows:

$$u = \frac{F}{k} \quad [1.57]$$

where the applied normal force is 1e6 N, the normal stiffness is 1e9 N/m and the corresponding normal deformation is 1e − 3 m.

The simulation results of DICE2D are plotted in Figure 1.20. A vibration is observed in the beginning, which decreases with time due to the kinematic energy being absorbed by the dashpot. The numerical result after 0.05 s will match the corresponding analytical solution. From the results, the following suggestion is made. The DEM is a typical dynamic relaxation-based method; in the simulation of quasi-static problems, the initial vibration must be addressed properly. For example the internal force at the beginning might be higher than the material strength, whereas the final equilibrium value is still low compared with the material strength. Because the material parameter is usually obtained from the physical test that satisfies the quasi-static equilibrium condition, directly applying the failure model during the entire calculation will result in unrealistic simulation results (e.g. model failures at the beginning vibration stage). In DEM simulation, one solution is to apply the boundary condition in a gradual manner.

In this example, only elastic deformation of the spring is considered for several reasons. For the benchmark example, rather than directly running a complex problem that includes all aspects of the DEM, it is always helpful to run a simple example that involves only a specific function of the code. This step-by-step method is useful for debugging a computer code. From this example, we can conclude that the normal spring and viscous dashpot are properly implemented in DICE2D. The normal spring tested in this example is in a bond form made up from two subsprings (shown in Figure 1.11). Because there is no moment produced between two particles under pure tension force, the bond thickness ratio (WRT), the ratio between d^b and the minimal particle diameter of the bond, has no influence on the simulation result. However, the WRT would have significant influence when shear loading is considered.

1.5.4. *Elastic deformation of spring under shear*

The two-particle system of the earlier example is tested under shear loading of 1e6 N. Under the shear force, the elastic deformation can be obtained from equation [1.57]. However, because the shear force will produce a moment on the top particle, a bond spring contact is required to balance this moment. Moreover, the boundary conditions of the bottom particle have to fix the movement along the x-direction and rotation as well. The model parameters of this example are listed in Table 1.5. Unlike the uniaxial tension case, the normal stiffness is set to a large value to get rid of the influence of normal stiffness on the simulation result. Because the normal stiffness will contribute to the bending of the contact, a large value is needed to filter out the bending influence on the final shear deformation. In this example, instead of using viscous damping, the automatic local damping is adopted to check the implementation. The expected shear deformation for the problem is 1e − 3 m. It should be mentioned that all the failure-related parameters have been set as large enough. Therefore, the problem is in an elastic (or intact) range.

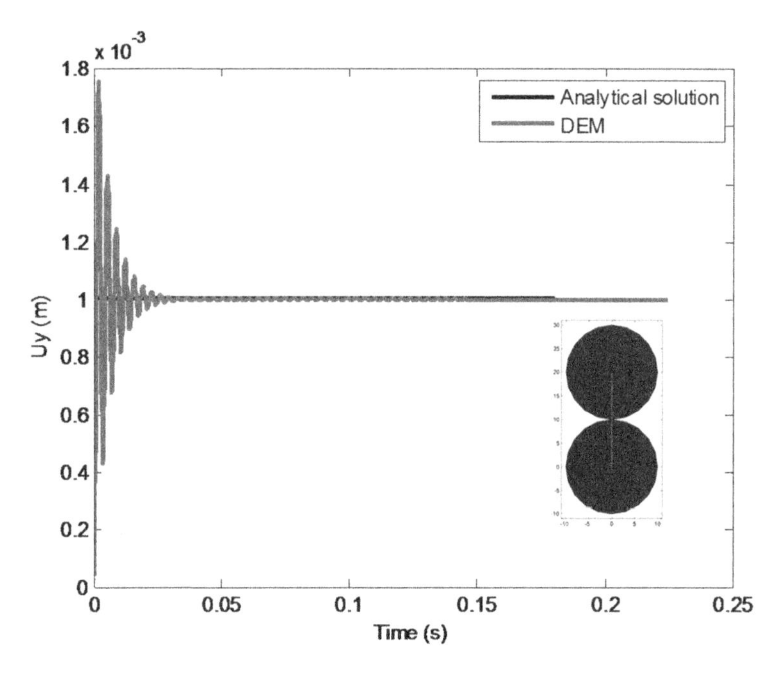

Figure 1.20. *Analytical and numerical results of the direct tension test of two particles*

Number of particles	2	Normal viscous coefficient (s)	0.0
Mean particle size (m)	10	Local damping	0.8
Density (kg/m^3)	1	Time step reduction factor	0.2
Normal stiffness (N/m)	500e9	Total steps	6,000
Shear stiffness (N/m)	1e9	Gravitational acceleration (m/s^2)	0

Table 1.5. *Model parameters of the shear test*

First, the WRT is set to zero. The simulation result is shown in Figure 1.21(a). It can be found that even with a very stiff normal stiffness (two particles are locked in the normal direction) the system is still unstable under shear loading. The upper particle will rotate around the fixed particle. The displacement of the upper particle cannot achieve a stable (equilibrium) value. Just keep in mind that the system is still intact; however, it is unstable. From this perspective, the basic element of the classical DEM is shear unstable, which is not the case of the real physical model. Therefore, the classical DEM might be unsuitable for modeling continuous problems. To overcome this shortcoming, the bond contact can be used. Figure 1.21(b) shows the simulation result where WRT is 0.8. It can be observed that a stable solution can be obtained, which is also close to the expected result. The difference between the predicted value and the analytical solution is due to the contribution from bending not being completely removed in the DEM simulation because the normal spring is not perfectly rigid. This example shows the importance of bond contact. It also verifies the implementation of the bond model and automatic damping in DICE2D.

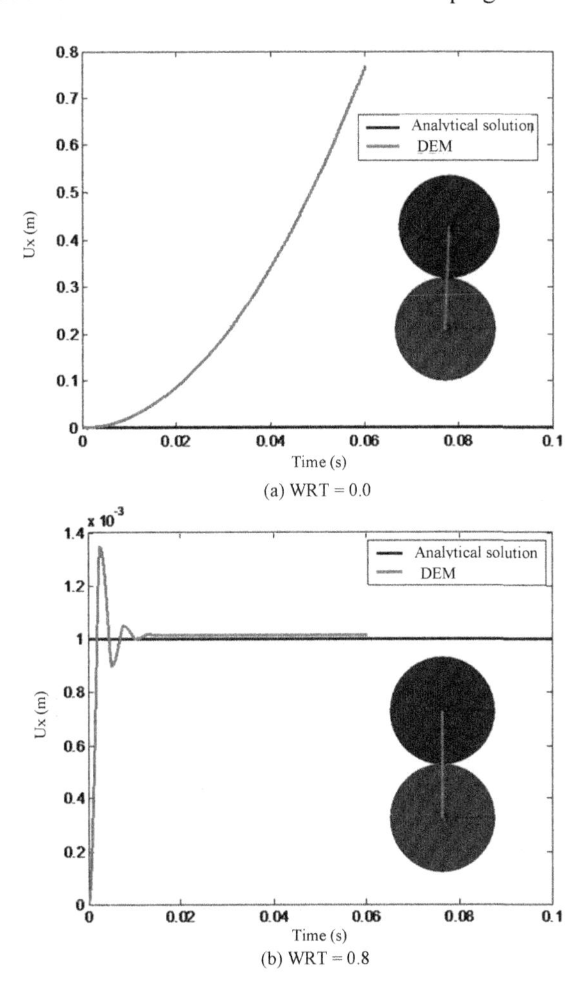

Figure 1.21. *Analytical and numerical results of the direct shear test of two particles under different bond ratios (WRT = 0.0 and 0.8)*

The moment transfer between particles is also reflected as an interlocking effect (e.g. [KAZ 10]). This can be simulated using the polygon particles [KAZ 10] or the irregular clumps made up from many circular particles [CHO 07]. However, the implementation of polygon particles is complex and computationally costly. The clump scheme will require additional time to prepare a clump particle model. Moreover, the clump scheme reduces the actual resolution of the computational model into the clump size. The bond model used in DICE2D is able to represent the moment transfer between particles in a concise way. There is only one dimensionless geometric parameter WRT introduced in the model.

1.5.5. *Failure of normal spring under tension*

One major advantage of the DEM is the ability to model the progressive failure of a material. The basic principle of the fracturing simulation using the DEM is to replace the complex fracturing and fragmentation process with a number of elemental failure events at the spring bond level. Tensile failure of a material is mainly controlled by tension failure of the normal springs. Therefore, the implementation of a tensile failure model in the DEM code is essential to reasonably correct simulate fracturing problems. In this section, the tension failure of a single normal spring is simulated by using DICE2D. The computational model will use the same model as in the previous example. The model parameters are shown in Table 1.6. In this example, a vertical velocity of 0.0001 m/s is applied to the upper particle. The bottom particle is fixed in the x-, y- and θ-directions. Instead of using viscous damping, a local damping of 0.8 is adopted to simulate a quasi-static loading condition. From the strain energy variation of the system (Figure 1.22(a)), it can be concluded that the system is energy stable under velocity loading. Moreover, the corresponding analytical displacement–force curve of the problem is a triangle that can be directly obtained. Figure 1.22(b) shows the comparison between the numerical and analytical predictions. A close fit is obtained. It can be concluded that the implementation of tension failure in DICE2D is correct.

Number of particles	2	Tension strength (N)	2e4
Mean particle size (m)	10	Local damping	0.8
Density (kg/m^3)	1	Time step reduction factor	0.2
Normal stiffness (N/m)	1e9	Total steps	2,000
Shear stiffness (N/m)	1e9	Gravitational acceleration (m/s^2)	0

Table 1.6. *Model parameters of the direct tension test*

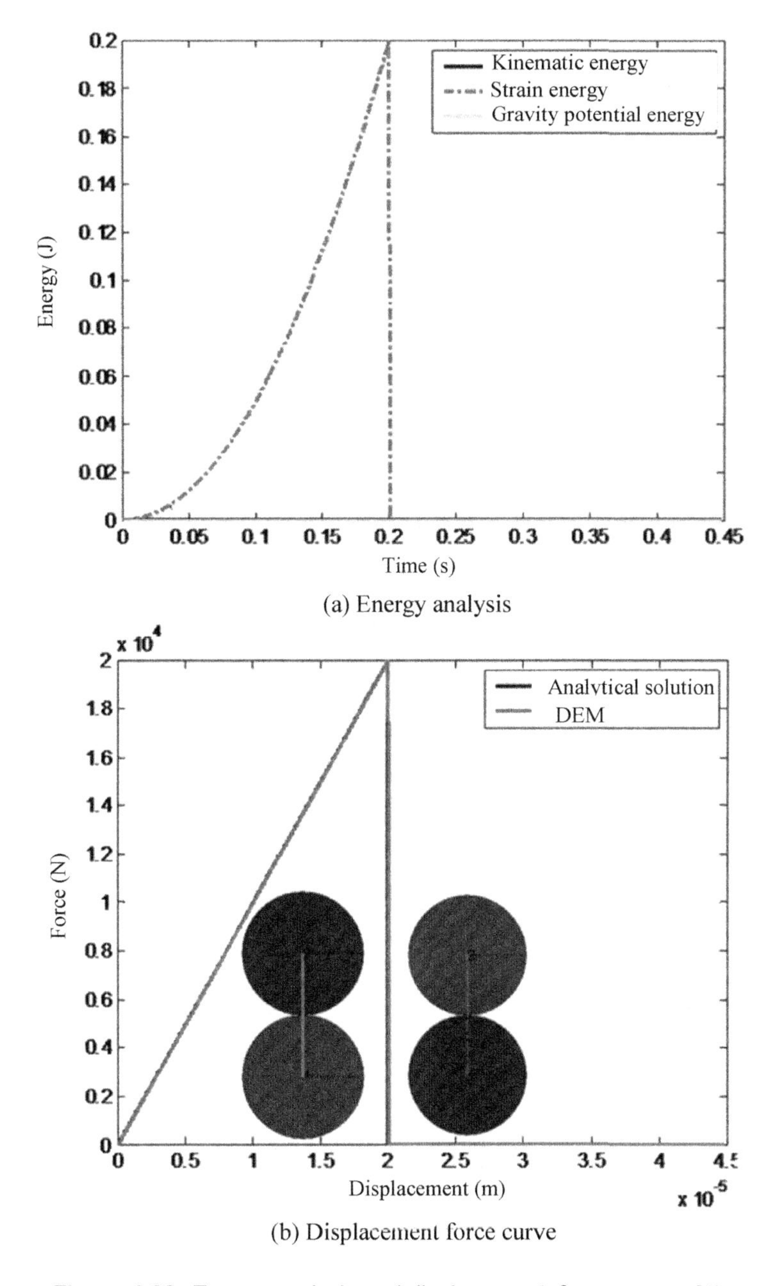

(a) Energy analysis

(b) Displacement force curve

Figure 1.22. *Energy analysis and displacement–force curve of the uniaxial tension problem when failure is considered*

The bond contact model before and after failure is also shown in Figure 1.22(b). The contact pair is a bond contact before the failure, represented as a solid line. After the bond contact is broken under tension, it will turn into a normal contact (dashed line). From energy analysis, it is found that the strain energy will be first increased to the peak value and then totally released (see Figure 1.22(a)).

1.5.6. *Failure of shear spring under shear*

To check the implementation of the Mohr–Coulomb failure criterion of the P2P contact, shear failure of the two-particle system, as shown in Figure 1.23, is simulated using DICE2D. The loading conditions of the bottom particle are kept the same. For the upper particle, a normal force of 1e5 N is applied vertically. In the meantime, a horizontal velocity of 1e − 4 m/s is applied to the upper particle, while its rotation is fixed. The model parameters are listed in Table 1.7. Except for the elastic and tension strengths, two additional parameters, cohesion and friction angle, are involved in the calculation. For this example, the shear force will increase gradually under the shear velocity until it reaches the maximum shear strength of the bond. The peak value can be easily calculated as $F_n \tan(\phi) + C$. After the bond is broken, the shear force will not change with the shear displacement but rather keep a constant value as $F_n \tan(\phi)$. It should be mentioned that because the normal force is applied at the beginning, an initial vibration of the strain energy is observed. As shown in Figure 1.24(a), after the vibration (caused by the normal force), the strain energy will increase gradually until it reaches the peak value. From the beginning to the peak, the bond is in an elastic range. When the bond breaks, the strain energy will be partially released. Unlike the tensile failure, the energy released from the shear failure is much less. Figure 1.24(b) shows the displacement–force curve predicted using DICE2D and the corresponding analytical solution. It can be concluded that the Mohr–Coulomb model is properly implemented in DICE2D.

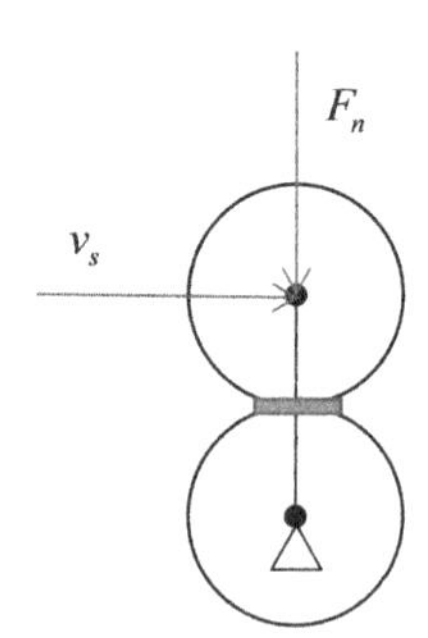

Figure 1.23. *The computational model of shear failure of the two-particle system*

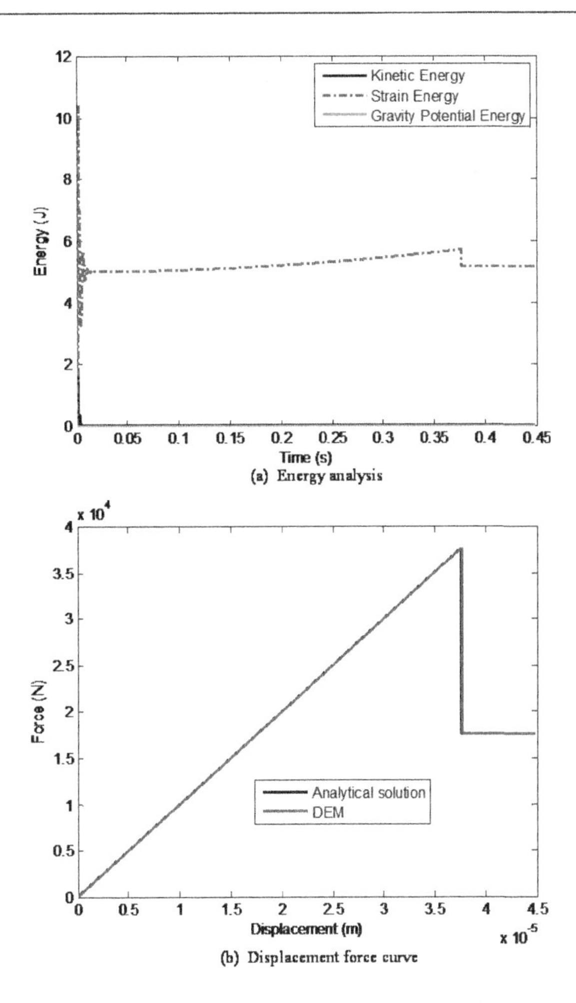

Figure 1.24. *Energy analysis and displacement–force curve of direct shear test of the two-particle system*

Number of particles	2	Cohesion (N)	2e4
Mean particle size (m)	10	Friction angle	10
Density (kg/m^3)	1	Local damping	0.8
Normal stiffness (N/m)	1e9	Time step reduction factor	0.2
Shear stiffness (N/m)	1e9	Total steps	2,000
Tension strength (N)	2e4	Gravitational acceleration (m/s^2)	0

Table 1.7. *Model parameters of the direct shear test*

1.5.7. *Newton balls*

Many sophisticated contact detection and treatment models have been developed for FEMs, for example the augmented Lagrangian method [SIM 92]. However, contact detection is still considered by many researchers as the most distinct feature of the DEM. I think that the main difference between FEM and DEM should be that the results are totally controlled by the contacts in the DEM, whereas for FEM, the continuum material constitutive model plays an important role. Therefore, some mesh-enriched DEMs, for example FEM/DEM [MUN 04] and Universal Distinct Element Code (UDEC) [ZHA 08], are not regarded as pure DEMs. Because contacts are the dominant elements of a DEM code, debugging the implementation of contact detection becomes the most critical part. In DICE2D, there are two types of contact detection: P2P and W2P. This example focuses on the P2P contact detection.

Here, the Newton balls are selected as the benchmark example. As shown in Figure 1.25, there are two particles linked to two strings. The model parameters are given in Table 1.8. The right ball is assigned with an initial velocity of 30 m/s. The ball will move according to the governing equation of the pendulum problem. It will first move to the peak point and then turn back to hit the left ball. The right ball will stop and transfer all kinematic energy to the left ball. This right-to-left cycle will continue forever if there is no energy consumption in the system.

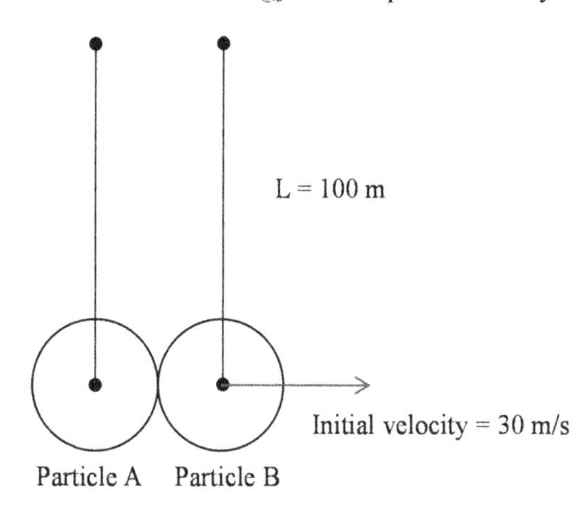

Figure 1.25. *Computational model of the Newton balls*

Figure 1.26 shows the simulation results using DICE2D. It can be concluded that the contact detection and treatment of the P2P contact have been properly implemented. The Newton balls can be viewed as a combination of the pendulum problem and an elastic collision problem. Energy analysis of the problem is shown in Figure 1.27(a). There are three energies: strain energy, kinematic energy and gravity potential energy. The energy balance happens mainly between the kinematic energy and the gravity potential energy. For the strain energy, the exchange between the strain energy and the kinematic energy happens at the collision for a very short time, and after collision, the gravity potential energy is not exactly zero. This slight lag caused the sum of the strain and the kinematic energies at two collision points to be different. However, from the energy analysis beyond the collision points, energy equilibrium is still satisfied. Therefore, the implementation of the P2P contact treatment in DICE2D is still correct. Moreover, to have a quantitative comparison, the corresponding analytical solution of the positions of the two balls is obtained using equation [1.56] and compared with the numerical results (Figure 1.27(b)).

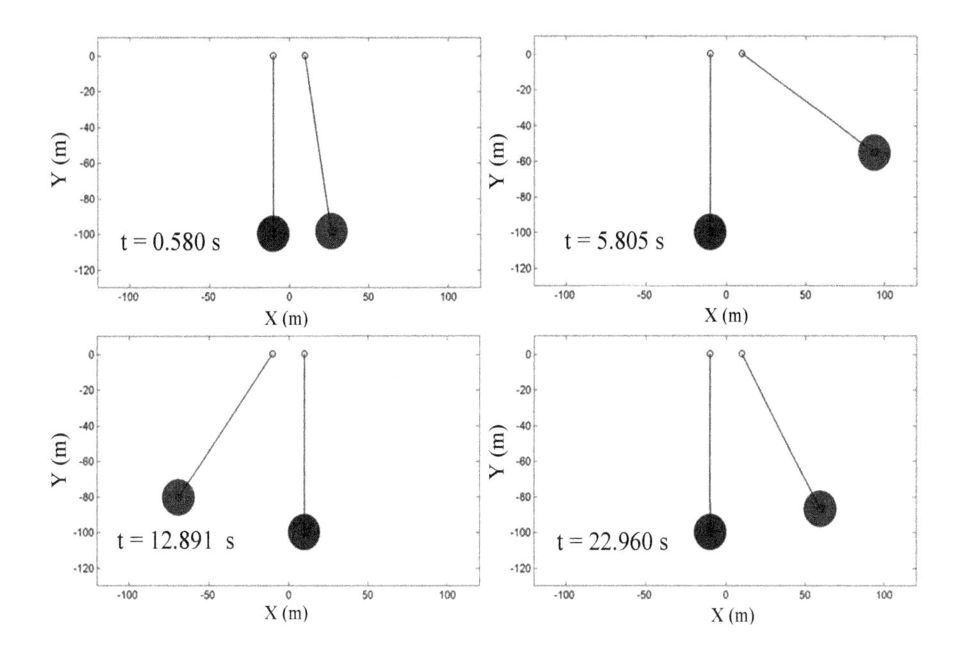

Figure 1.26. *The Newton balls simulated using DICE2D*

Number of particles	2	Cohesion (N)	0
Mean particle size (m)	10	Friction angle	0
Density (kg/m^3)	1	Local damping	0.0
Normal stiffness (N/m)	1e7	Time step reduction factor	0.5
Shear stiffness (N/m)	1e7	Total steps	5,000
Tension strength (N)	0	Gravitational acceleration (m/s^2)	10

Table 1.8. *Model parameters of the Newton balls*

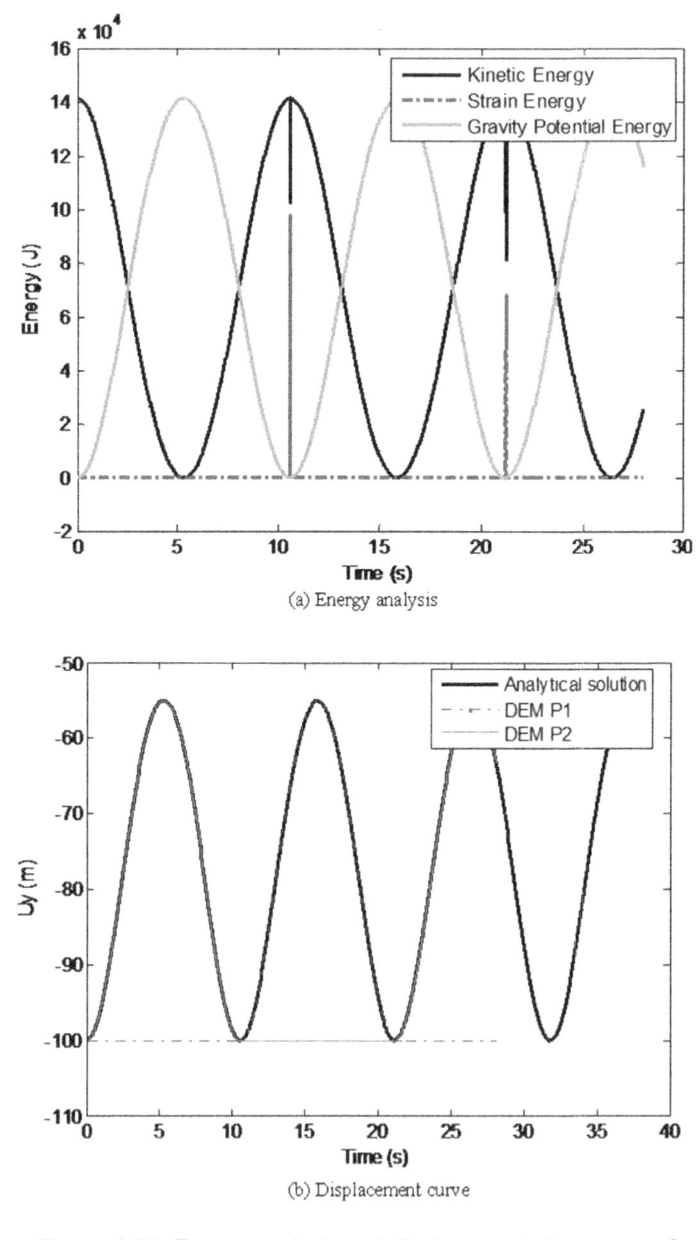

(a) Energy analysis

(b) Displacement curve

Figure 1.27. *Energy analysis and displacement–time curve of the Newton balls predicted using DICE2D*

1.5.8. *Bounce back ball*

In this example, the implementation of the W2P contact is checked. The computational model is shown in Figure 1.28. It is a modified version of the falling ball problem. The ball will first be assigned with an initial velocity of 30 m/s. It will first fly to the peak point and then fall down under the effects of gravity, eventually bouncing back again. This bounce back will repeat forever if only the elastic collision is considered. When the viscous–elastic collision is considered, the system energy will decrease with time and the ball will finally settle on the wall. The model parameters are listed in Table 1.9. The tension, friction and cohesion of the wall are set to zero. Two cases are modeled: the elastic collision (viscous coefficient is zero) and the viscous–elastic collision (viscous coefficient = 1e − 4 s). Figure 1.29 shows the ball position at different times predicted using DICE2D. The ball will bounce back when it hits the wall; therefore, the wall element is properly implemented. Similar to the Newton balls problem, energy analysis is conducted (Figure 1.29(a)). It can be concluded that the energy balance is satisfied for the P2W contact in DICE2D.

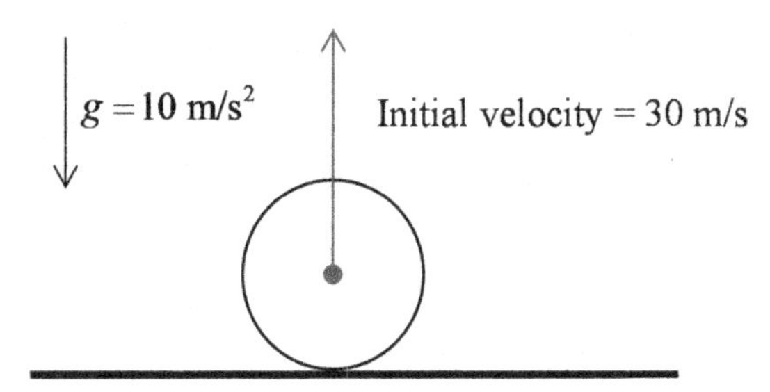

Figure 1.28. *Computational model of the bounce back ball problem*

Number of particles	1	Viscous coefficient	0/1e − 4
Mean particle size (m)	10	Friction angle	0
Density (kg/m^3)	1	Local damping	0.0
Normal stiffness (N/m)	1e7	Time step reduction factor	0.5
Shear stiffness (N/m)	1e7	Total steps	3,200
Tension strength (N)	0	Gravitational acceleration (m/s^2)	10

Table 1.9. *Model parameters of the bounce back ball problem*

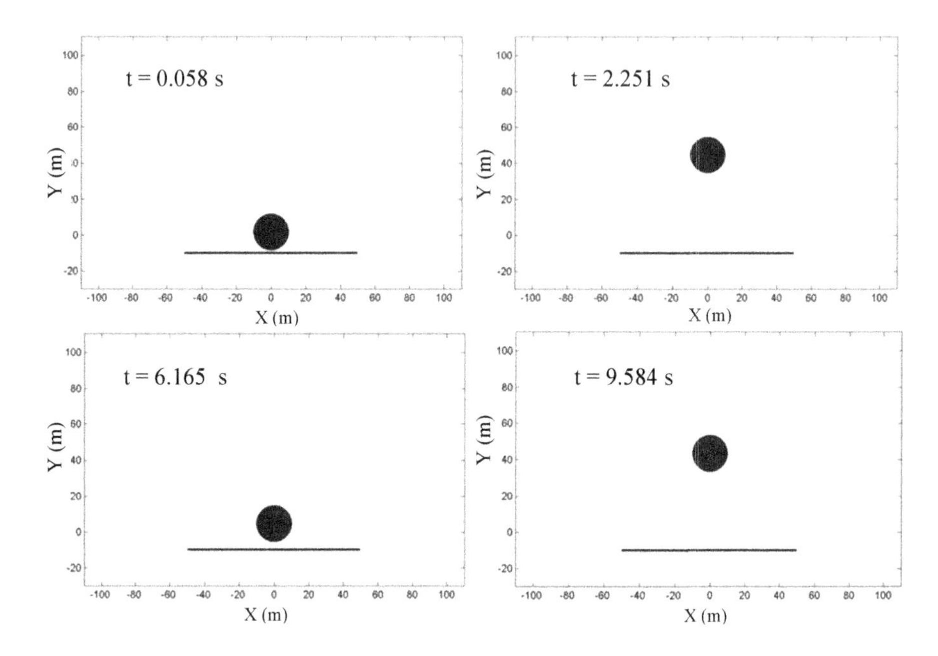

Figure 1.29. *Bounce back of the ball predicted using DICE2D*

The analytical solution and numerical prediction of the history of the ball's position with respect to time are shown in Figure 1.29(b). An exact match is obtained. In practice, energy will be consumed because of the breakage of contact points or the viscosity from fluid attached in the contact surface; therefore, the viscouselastic contact should be used for these situations. To further investigate the influence of viscous–elastic contact, the energy and displacement of the viscous–elastic collision case predicted using DICE2D are shown in Figure 1.30. The kinematic energy and its peak point of the system will decrease with the increasing number of collisions. From Figure 1.30(b), it can be observed that the frequency of the system will increase with the collision time. Therefore, the system will achieve a stable condition in an accelerated way. Imagining the actual scheme in real life, for example the bounce back of a ping-pong ball on a rigid floor, the simulation results are reasonable (Figure 1.31).

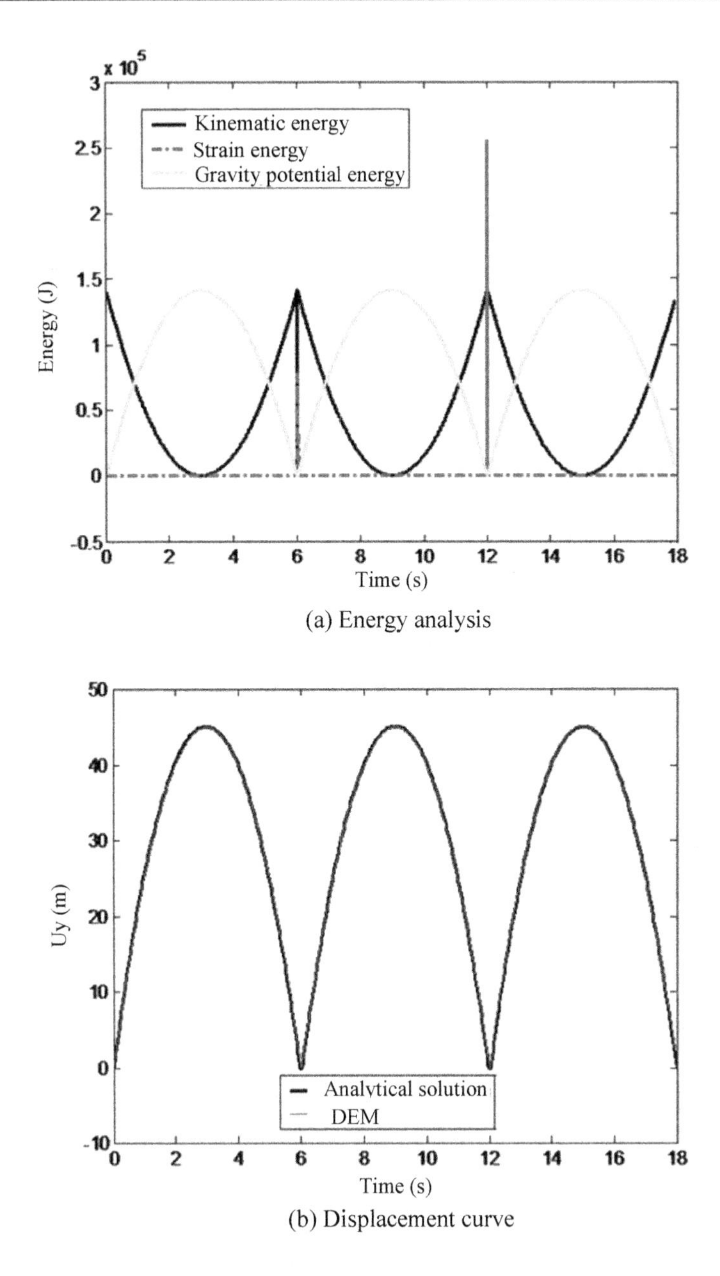

(a) Energy analysis

(b) Displacement curve

Figure 1.30. *Energy analysis and displacement curve of the bounce back ball problem (elastic collision)*

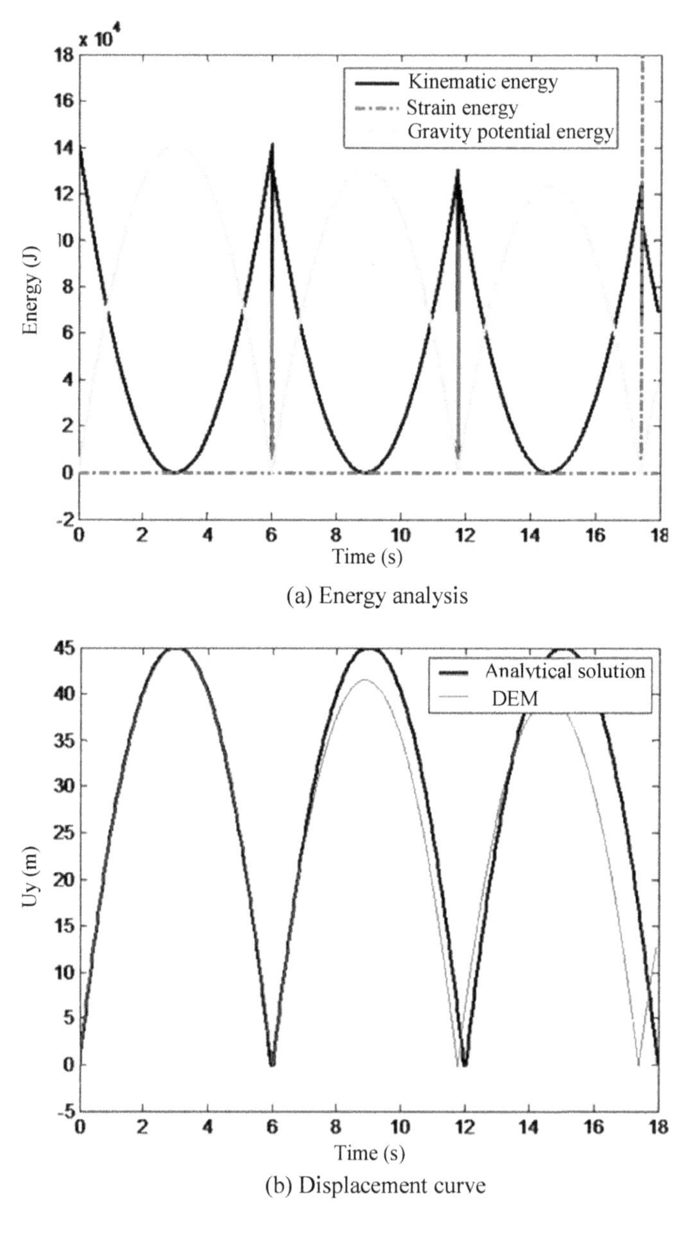

Figure 1.31. *Energy analysis and displacement curve of the bounce back ball problem (viscous–elastic collision)*

1.5.9. *Sliding particle*

In this section, the implementation of the W2P constitutive model in DICE2D is checked. The sliding particle problem is selected as the benchmark example (Figure 1.32). For the particle sliding along the inclined plane problem, the main interaction is from the friction force and the gravity subforce along the shear direction. The bounce back ball problem verified the implementation of the normal spring of the W2P contact. This example will further check the implementation of the shear spring and friction law of the W2P contact.

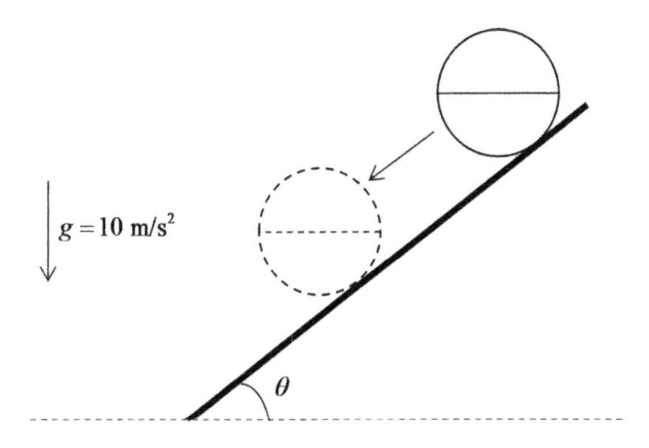

Figure 1.32. *Sliding particle problem*

The analytical solution of this problem is given as follows:

$$L(t) = \frac{1}{2} t^2 g (\sin\theta - \tan\phi \cos\theta) \qquad [1.58]$$

where L is the sliding distance, t is the time, g is the gravitational acceleration, θ is the inclination angle between the wall and the ground (30°) and $\o$ is the friction angle of the wall (20°). Table 1.10 shows the model parameters used in DICE2D. To satisfy the sliding condition described in equation [1.58], the rotation DOF of the particle is fixed during calculation.

Number of particles	1	Local damping	0
Mean particle size (m)	10	Wall tension	0
Density (kg/m^3)	1,000	Wall friction	20
Normal stiffness (N/m)	1e8	Wall cohesion	0
Shear stiffness (N/m)	1e8	Time step reduction factor	0.1
Tension strength (N)	0	Total steps	800
Cohesion	0	Gravitational acceleration (m/s^2)	10
Friction angle	0		

Table 1.10. *Model parameters of the sliding particle problem*

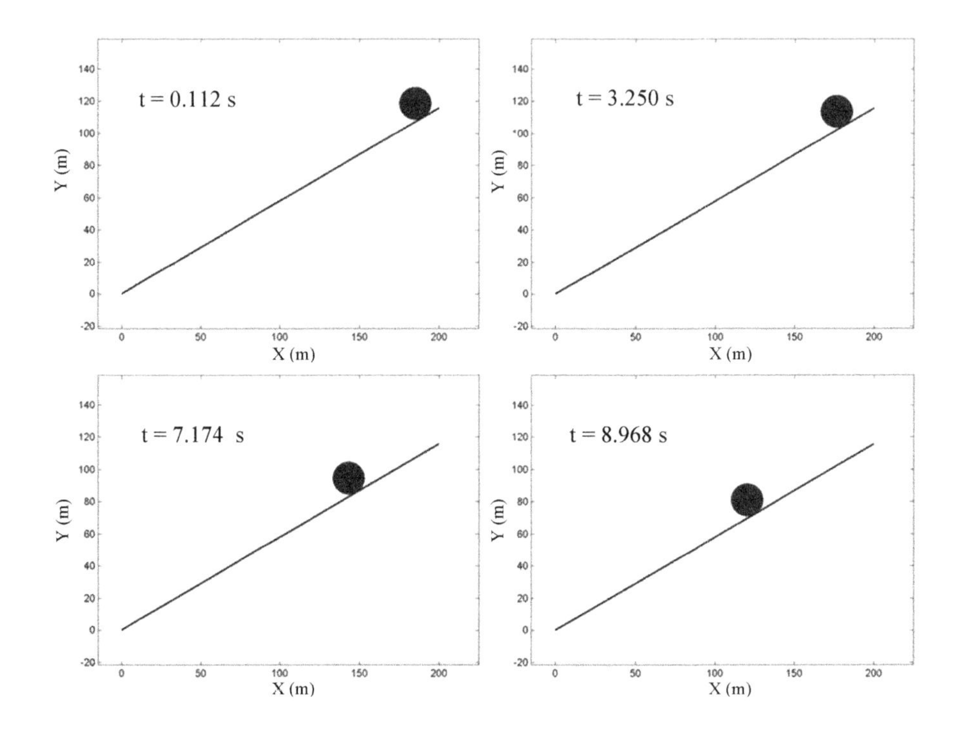

Figure 1.33. *Sliding particle problem predicted using DICE2D*

Figure 1.33 shows the sliding process of the particle predicted using DICE2D. The particle slides from the top of the plate to the bottom. Because the sliding distance is larger than the contact detection threshold value (0.4 particle radius), the P2W contact is active during the calculation. Therefore, this example also checked the implementation of W2P dynamic contact detection. Figure 1.34(a) shows the change of energies during the particle sliding. The figure shows that the kinetic energy of the system increases whereas the gravity potential energy decreases. However, energy balance is not observed because the friction absorbed some energy. This is not the due to the DEM simulation but due to the inherent nature of the problem. This can be further confirmed from the comparison between the numerical prediction on the sliding distance and the analytical prediction (Figure 1.34(b)).

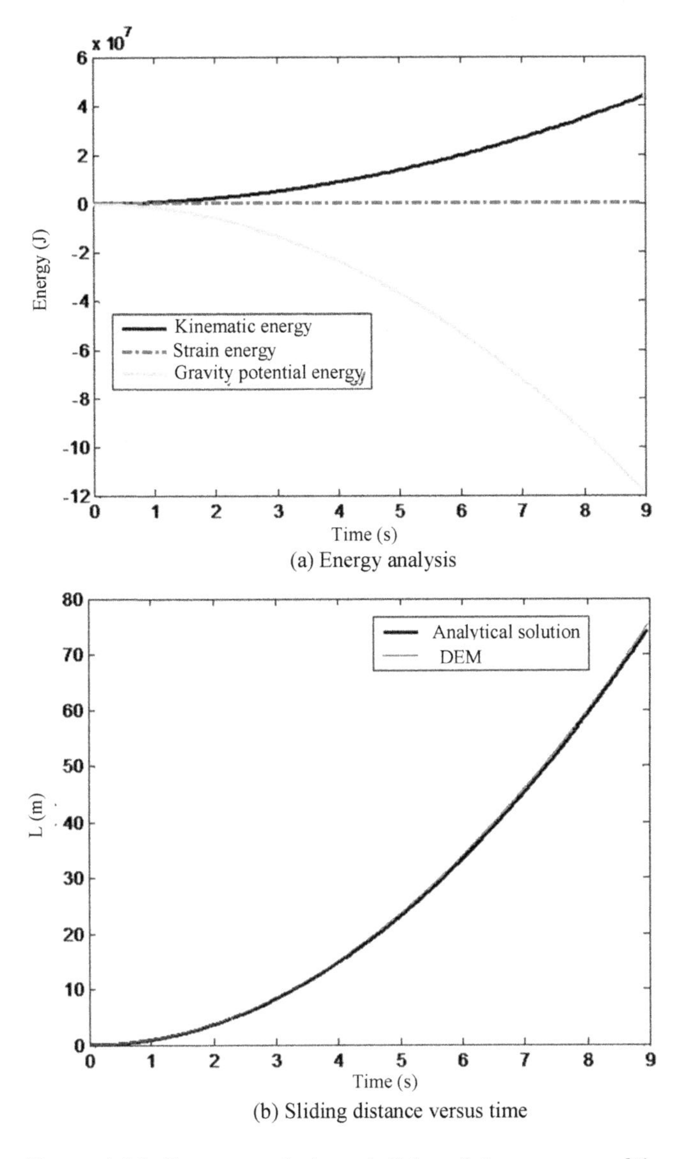

Figure 1.34. *Energy analysis and sliding distance curve of the sliding particle problem (rotation is fixed)*

1.5.10. *Sliding particle with rolling*

Until now, rotation of a particle has still not been checked. In this example, the particle rotation is verified using the particle sliding example. The computational model is the same as in the previous example. The only difference is that particle rotation is not fixed during calculation. The model parameters are also the same as those listed in Table 1.10. The only difference is that the friction angle between the wall and the particle is set to 80° to ensure the non-slipping condition. Under this condition, the analytical solution for the particle velocity can be given as follows:

$$v_L = \sqrt{\frac{4}{3}gh} \qquad [1.59]$$

where v_L is the sliding velocity along the wall, h is the current sliding height of the particle and g is the gravitational acceleration. Figure 1.35 shows the rolling and sliding of the particle. Even though a very large friction angle (80°) is adopted, the particle can still slide smoothly. Moreover, from the energy analysis (Figure 1.36(a)), it is found that the kinetic energy is nearly equal to the dissipated gravity potential energy. It can be concluded that the friction between the wall and the particle did not consume as much energy as in the previous sliding particle example. From Figure 1.35, the sliding distance is also much longer. From this example, it can be concluded that the friction angle used in the DEM cannot directly be used to represent the macroscopic observed friction angle in a physical test. A calibration is needed. Figure 1.36(b) shows the analytical solution and numerical prediction using DICE2D.

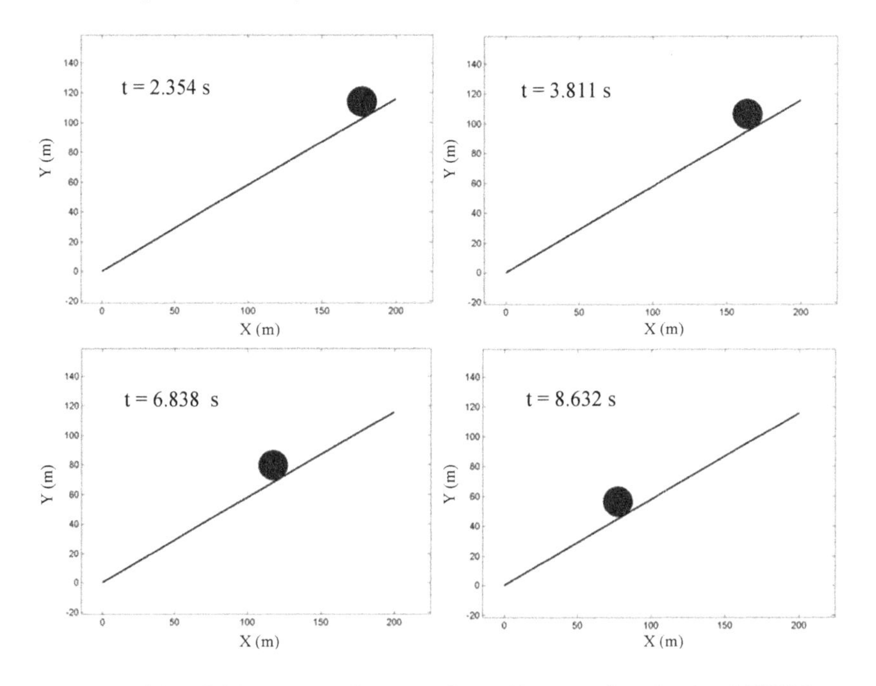

Figure 1.35. *Sliding and rolling particle problem predicted using DICE2D*

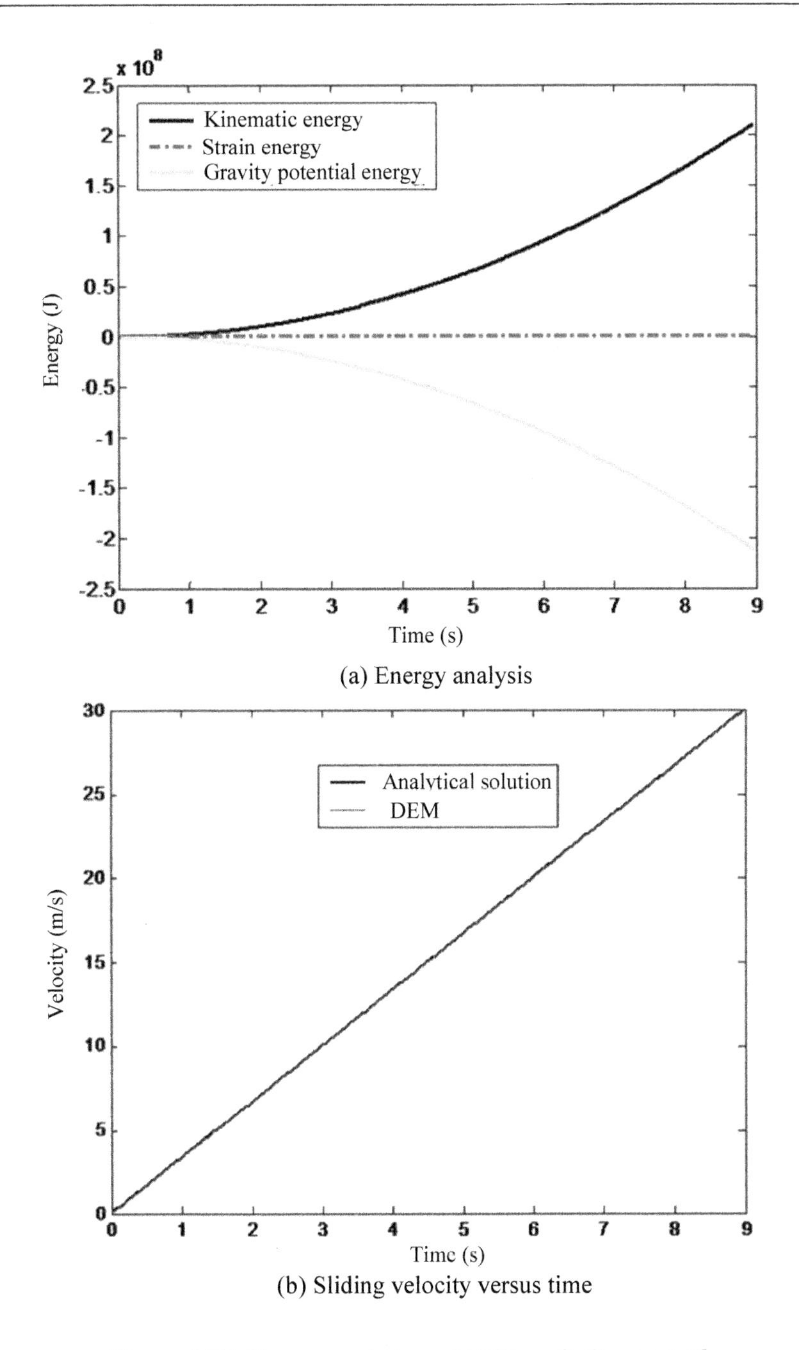

(a) Energy analysis

(b) Sliding velocity versus time

Figure 1.36. *Energy analysis and sliding velocity curve of the sliding and rolling particle problem*

Until now, all basic functions of DICE2D, for example particle movement with three DOFs and contact detection treatment of P2P, have been verified. Because these problems are very simple (maximum of two particles) and the analytical solutions are available, they can be used to debug the DEM code. In the following section, two more complex examples are used to further test DICE2D.

1.5.11. *Rock fall problem*

Rock fall is a typical hazard in mountainous areas. For example a rock boulder rolling and sliding from a slope will put any buildings below it in jeopardy. In this example, a rock fall problem is simulated using DICE2D. The computational model is shown in Figure 1.37. A high-rise building is simulated using 11 particles bonded together. To the left of the building, there is a slope with a large rock boulder. The boulder will slide down and hit the building, which may collapse. The model parameters are shown in Table 1.11. The strength of the building is set to zero to trigger the dynamic contact detection between different particles, as in a granular flow situation. Viscous–elastic collision is considered to model the energy consummation during the building collapse.

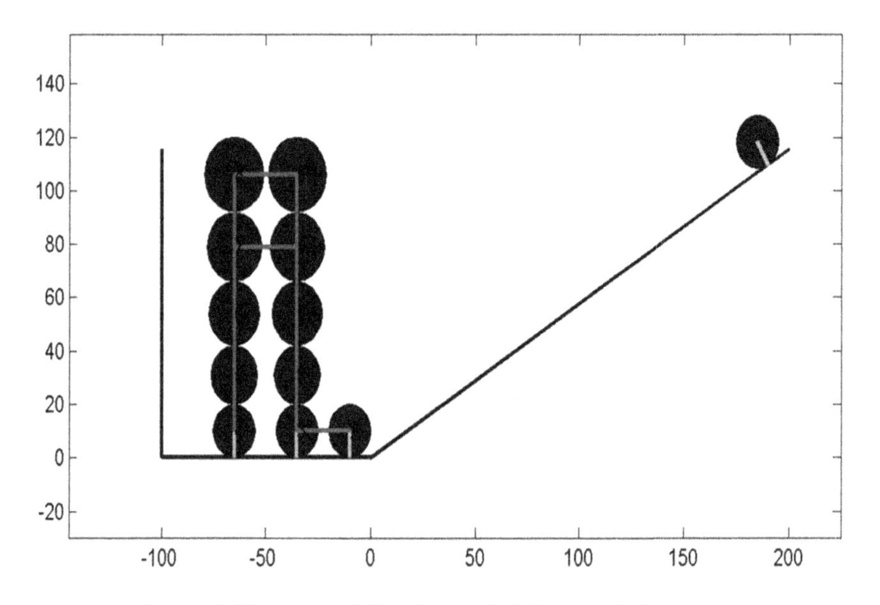

Figure 1.37. *Computational model of the rock fall problem*

Number of particles	12	Normal viscous coefficient (s)	1e − 2
Mean particle size (m)	11.66	Local damping	0
Density (kg/m^3)	1,000	Wall tension	0
Normal stiffness (N/m)	1e8	Wall friction	30
Shear stiffness (N/m)	1e8	Wall cohesion	0
Tension strength (N)	0	Time step reduction factor	0.1
Cohesion	0	Total steps	800
Friction angle	0	Gravitational acceleration (m/s^2)	10
Normal viscous coefficient (s)	1e − 2		

Table 1.11. *Model parameters of the rock fall problem*

The failure process of the building is shown in Figure 1.38. It can be observed that the contact between the particles and the wall are dynamically formed and properly treated. This example shows that these functions, separately verified in the aforementioned examples, can work together properly to handle a complex problem. Moreover, the ability of DICE2D to address a computational model with different size particles is also confirmed.

1.5.12. *Beam collision*

The ability of the DEM to model fragmentation is regarded as one of its most attractive advantages in rock mechanics. For a classical continuum-based method, for example FEM, to model a transaction from continuum to discontinuum, an additional element (contact element) and a failure model (linear fracture mechanics-based model) are needed. The DEM is a natural process and has no need of any further modification. In this example, DICE2D is used to model a fragmentation problem involving large deformation and contact collision. Figure 1.39 shows the computational models. There are two particle models: one is a square loose packing model and the other is a triangular dense packing model. The purpose is to show the influence of particle packing on the simulation results (Figure 1.40). The model parameters are shown in Table 1.12. There are 100 and 98 particles for the square- and triangular-packed models, respectively. All other parameters of these two models are the same.

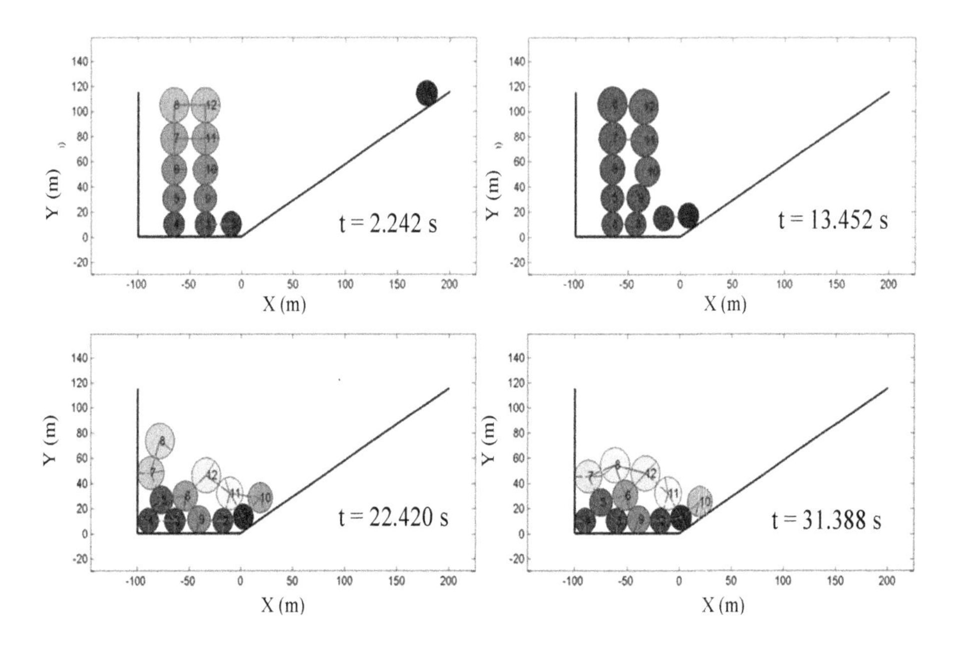

Figure 1.38. *Dynamic collapse process of the building modeled using DICE2D. For a color version of the figure, see www.iste.co.uk/zhao/computing.zip*

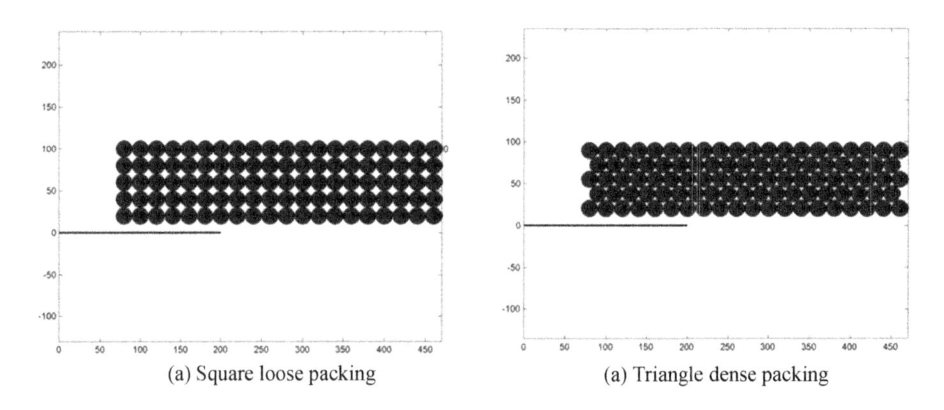

(a) Square loose packing (a) Triangle dense packing

Figure 1.39. *Beam collision problem using different particle packing*

Number of particles	100/98	Normal viscous coefficient (s)	0
Mean particle size (m)	10	Local damping	0
Density (kg/m^3)	1,000	Wall tension	0
Normal stiffness (N/m)	1e8	Wall friction	80
Shear stiffness (N/m)	1e8	Wall cohesion	0
Tension strength (N)	1e9	Time step reduction factor	0.1
Cohesion	1e9	Total steps	800
Friction angle	80	Gravitational acceleration (m/s^2)	10
Normal viscous coefficient (s)	0		

Table 1.12. *Model parameters of the beam collision problem*

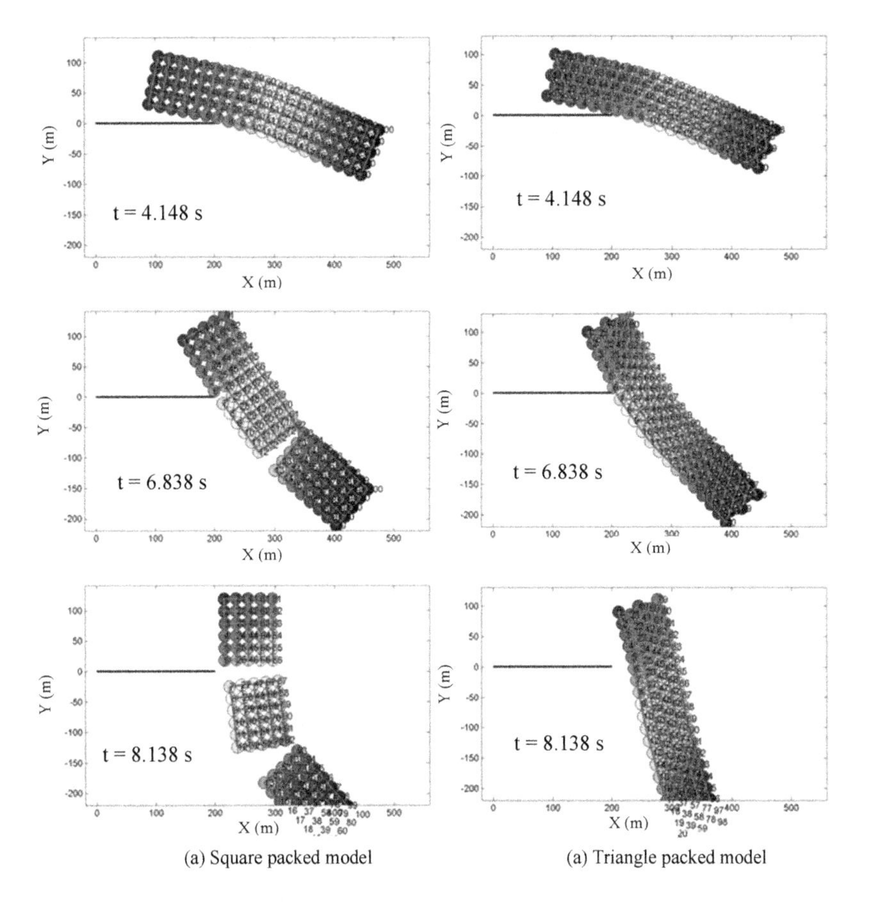

(a) Square packed model

(a) Triangle packed model

Figure 1.40. *Simulation results of the beam collision problem. For a color version of the figure, see www.iste.co.uk/zhao/computing.zip*

From all of the benchmark examples, the implementation of DICE2D was fully tested and verified. In any computer program, bugs are unavoidable. The only way to eliminate them is to run more examples. At this point, a serial DICE2D, which can be used as the DEM platform of parallelization, has been developed and is ready.

1.6. Conclusion

In this chapter, serial implementation of a DEM code (DICE2D) is described. The data structure design of particle, wall and boundary conditions, flowchart of key algorithms such as contact detection and contact treatment, and constitutive model of P2P and W2P contacts are presented. The target of DICE2D is to provide a DEM platform through a friendly programming environment for parallelization study. A few benchmark examples are provided to verify fundamental components of the DEM implementation. Two complex examples on rock fall and fragmentation are provided to show the ability of the developed code to model granular flow and fragmentation problems. Modeling results indicate that DICE2D is a properly developed DEM code and is ready to be parallelized.

2

Multi-core Implementation

In this chapter, DICE2D is parallelized for multi-core PCs. First, a brief introduction of parallel computing using multi-core PCs and its implementation using the Parallel Computing Toolbox® of MATLAB® is presented. Then, the performance of serial DICE2D is analyzed to aid the selection of essential sections that might be effective for parallelization. After this performance analysis, implementation details of parallel DICE2D are described. Finally, four examples are considered to verify the implementation.

2.1. Multi-core personal computer

The term "personal computer" refers to a general-purpose computer whose size and capabilities are small, and whose price is sufficiently low to make it available to individuals. A PC is also called a microcomputer, which indicates that its computing power is much less than that of a supercomputer. However, with the development of hardware and software technologies in computer science, modern PCs have become the dominant tool in performing scientific computing and numerical modeling. The primary reason for this widespread use is that the software and operating systems in PCs are relatively user-friendly. In addition, with the advancement of CPUs and memory used in modern PCs, some engineering problems can also be solved easily using a PC. Recently, a new term "personal high performance computing" (PHPC) [CHA 06] has been proposed. PHPC aims to solve problems that previously could be handled only by a supercomputer using a normal PC. This target may be attained in the near future if a 50-core CPU and a 128-bit operating system are developed sufficiently. Modern PCs equipped with advanced multi-core processors already provide adequate computing power and memory for scientific computing; for example, a laptop equipped with a 2.7-GHz CPU and 8-GB memory is sufficient to run a DEM with a few million particles. The typical structure of a quad-core processor is shown in Figure 2.1. The multi-core processor provides better performance, including multiple execution units that helps execution of the instructions per cycle separately in different cores; this approach has the advantage of simultaneously handling multiple tasks. In practice, the level of performance improvement using multi-core processors is highly dependent on the code used. Many typical DEM codes, however, do not consider the parallelization of multi-core PCs. Typically, the parallelization of a code on a multi-core PC is relatively simple because it manages only the shared-memory environment and does not need to consider the task distribution and communication between different processors. Parallel programming environments such as OpenMP [OPE 10], pThreads [DIC 96] and Threading Building Blocks (TBB) [TBB 10] can also be used to implement the multi-core version of an existing code; however, there are several shortcomings [RIC 08]. First, adjustments of the existing code are required for maximum utilization of the computing resources. Second, managing thermal issues is more difficult on multi-core designs than on single-core designs. Third, multi-thread code often requires the complex coordination of threads, which makes it difficult to locate bugs. The interaction between different threads can also lead to safety problems. However, from my experience with parallel computing using a multi-core processor, it can be a stable and promising solution for research purposes.

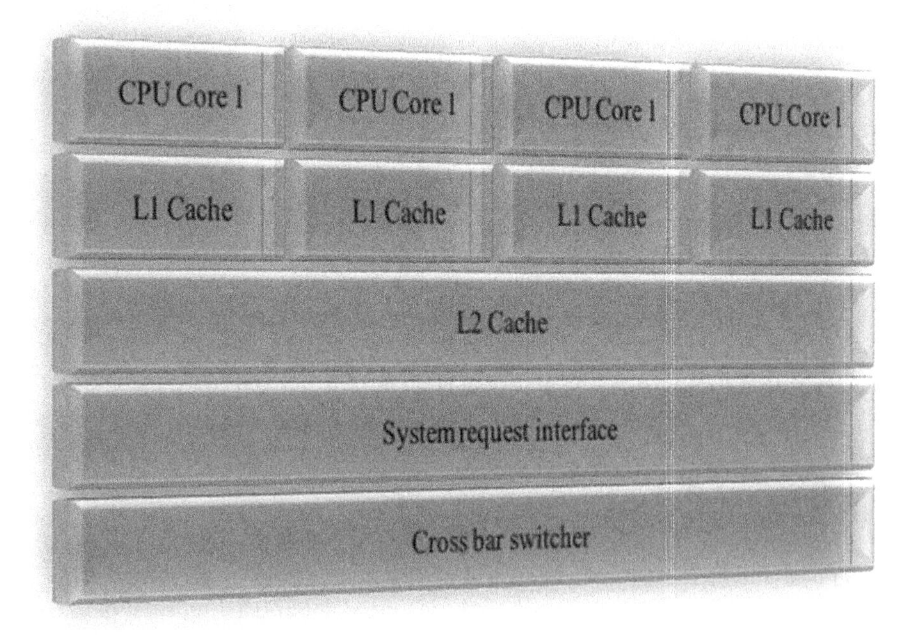

Figure 2.1. *Structure of a general quad-core processor*

2.2. Multi-core implementation using MATLAB

OpenMP is the most common technique used to parallelize computer code for multi-core PCs. However, MATLAB does not currently provide direct support for OpenMP. Fortunately, it provides an alternative high-level parallelization solution, the Parallel Computing Toolbox of MATLAB, which provides an integration solution for parallel computing, including multi-core PCs, GPU computers and clusters. A number of techniques provided in the Parallel Computing Toolbox can be adopted for the parallel implementation of DICE2D, for example the MPI, single program multiple data (SPMD), Distributing Arrays and *parfor*. In MATLAB, the MPI functions are wrapped to precisely control the data communication between different processors. However, parallelization using MPI involves computational task distribution and communication and is difficult in terms of implementation. The SPMD can run a block of code in parallel to process its own individual data separately. This technique is useful to process data processing tasks. There is no communication and interaction between different subdivided data sets, which is not true in the case of DEM. Distributing Arrays save memory and computing by distributing a huge matrix to different processors. In DEM, there are many small matrix operations rather than a single huge matrix operation. Therefore, Distributing Arrays are not adopted. The last choice, *parfor*, is a high-level parallel technique in which both memory distribution and communication are handled automatically by MATLAB. Figure 2.2 shows the strategy of *parfor* computing provided by MATLAB.

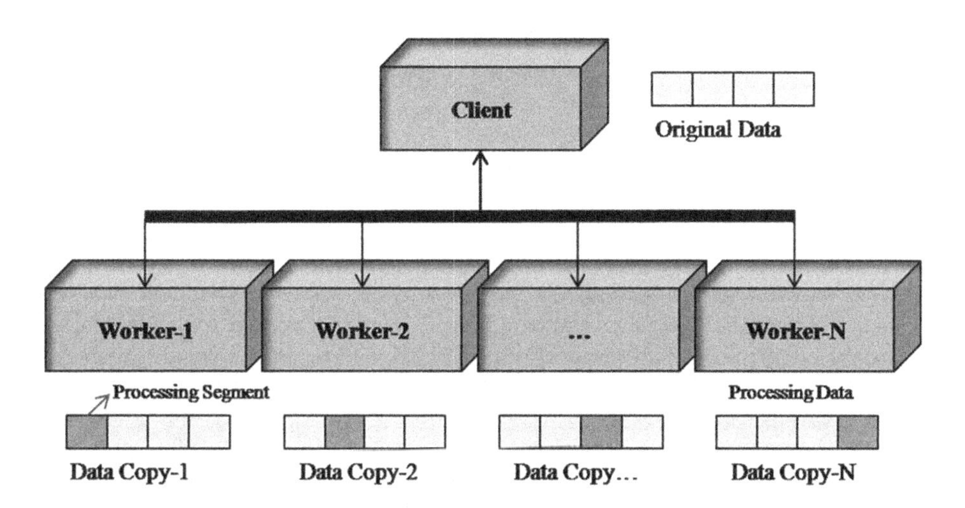

Figure 2.2. *Principle of parfor parallel computing strategy of MATLAB*

A worker–client mode is adopted. The client handles the main stream of computing, such as the serial code sections, job distribution and communication. When a loop is parallelized using *parfor*, all computation tasks are divided into a few subsections that are sent to workers for calculation. When computing is finished, the results are sent to the client. It should be noted that MATLAB sends the whole array to the workers rather than just part of it, whereas only the assigned parts are calculated by the corresponding workers. The workers can be regarded as virtually built computers with their own CPU and memory. Owing to data being communicated among the workers and the client, the performance of the parallel code might not always be faster than that of the serial code. For the above reasons, among the available techniques, *parfor* is selected for the parallel implementation of DICE2D. This chapter focuses on the parallelization of DICE2D in multi-core PCs using *parfor*.

The parallelization of code using *parfor* can easily be achieved by replacing the keyword *for* in the serial version with the keyword *parfor*. For those who are familiar with OpenMP, the usage of *parfor* can be simply viewed as *omp parallel for*. Example MATLAB codes are shown in Table 2.1.

Serial Code	Parallel Code

function Serial_T01
tic;
N=1e3;
for i=1:N
    A(i)=sin(i*2*pi/1024)
end
plot(A);
toc
	function Parallel_T01
	matlabpool open local 4
	tic;
	N=1e3;
	parfor i=1:N
	    A(i)=sin(i*2*pi/1024)
	end
	toc
	matlabpool close

Table 2.1. *MATLAB code using parfor*

The first column of the table is the serial code. The corresponding parallel version is shown in the second column. Timer functions, *tic* and *toc*, are used to record the computing time of these codes. The only difference that can be observed between the main body of these two codes is the keyword *for*. In MATLAB2010b, to run a parallel code, a pool of workers must be explicitly allocated using the *matlabpool* function. The number of workers is limited by the available physical cores and type of MATLAB license. For example, my license of MATLAB2011b allows me to allocate a maximum of four workers. My laptop is a quad-core computer. Therefore, a pool with four workers is created using “matlabpool open local 4”. When computing is finished, it is important to close the pool using "matlabpool close". It should be noted that the latest version of MATLAB supports default pool settings. Details can be found in the manual or online help of MATLAB. These documents are freely available on the Internet.

The computational times of these two codes given in Table 2.1 are shown in Table 2.2. When N is small, the computational time of the parallel code is even higher than that in the case of the serial code. This increase is due to the overhead time used for memory allocation and the fact that the demand imposed by communication of the parallel code negates the benefit obtained from fast computing operations through the subdividing strategy. However, when N is larger than a given value, the parallel code becomes faster than the serial code. The maximum speedup is approximately 2. From this example, it can be concluded that *parfor* is effective in MATLAB. In section 2.3, this technique is used to parallelize DICE2D.

N	Computational time (s)		Speedup
	Serial code	Parallel code	
1.00E + 03	0.00118	0.17672	0.00670
1.00E + 04	0.00962	0.18950	0.05074
1.00E + 05	0.09554	0.20491	0.46626
1.00E + 06	0.86391	0.65985	1.30925
1.00E + 07	5.54261	2.49368	2.22266
1.00E + 08	65.9267	28.6730	2.29925

Table 2.2. *Run time of parallel and serial code in MATLAB*

2.3. Performance analysis

For parallel implementation, it is essential to conduct a performance analysis on the serial code to find the most computational time-consuming parts for parallelization. In MATLAB, the performance analysis can simply be achieved using the *profile* command. For example, to test DICE2D, the code shown in Table 2.3 can be used.

```
profile on
D2D(1);
profile view
```

Table 2.3. *Performance analysis of DICE2D*

The uniaxial compression test of a cube is selected as an example for the performance test of DICE2D (Figure 2.3). Details on the model parameters and boundary conditions are presented in section 2.5.1. The particle model used in the performance test is 9 × 9. A report on the computational time is shown in Figure 2.4. During calculation, the particle model is plotted to produce animation. In the meantime, images are stored in the hard drive. By turning on these functions, real-time information about the DEM simulation can be obtained. However, from the profile analysis, it can be observed that nearly all of the computational time was spent on the plotting and output rather than on the DEM calculation. This information can be found in the graphic display of Total Time Plot (see Figure 2.4) in which the dark bar refers to the self-run time (the time spent on the corresponding function). In DICE2D, the plotting and file outputs are difficult to parallelize because they are system-provided functions. Moreover, for DEM simulations, graphic outputs of the whole-particle model are usually not necessary. In the second run, the plotting and output functions are disabled. Figure 2.5 shows the computational time of DICE2D without plotting and outputs. From the total time analysis, it can be found that the majority of computational time is spent on the DEM main code. In the parallel implementation, the functions to turn the plotting and outputs on or off were added in DICE2D as a result of profile analysis. In the following (sections 2.5.1 to 2.5.4), when DICE2D is tested for parallelization, the outputs and plotting are all turned off. For this book, separate runs were conducted to produce figures.

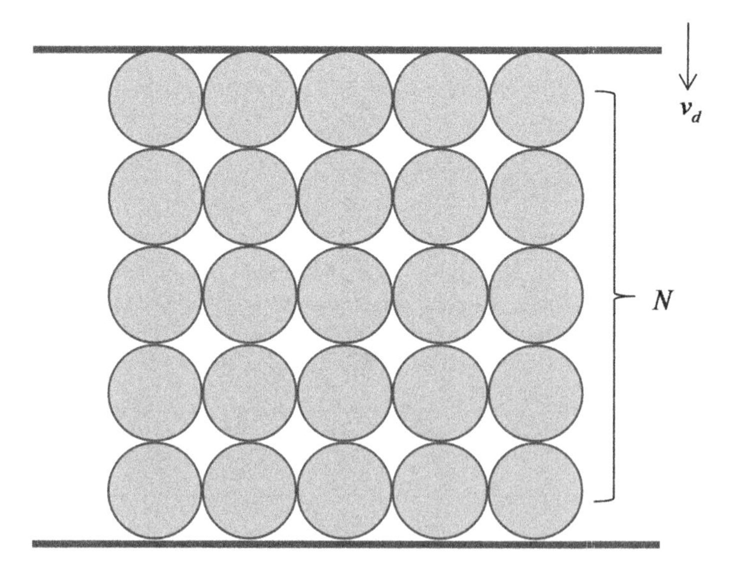

Figure 2.3. *Uniaxial compression test of a cube*

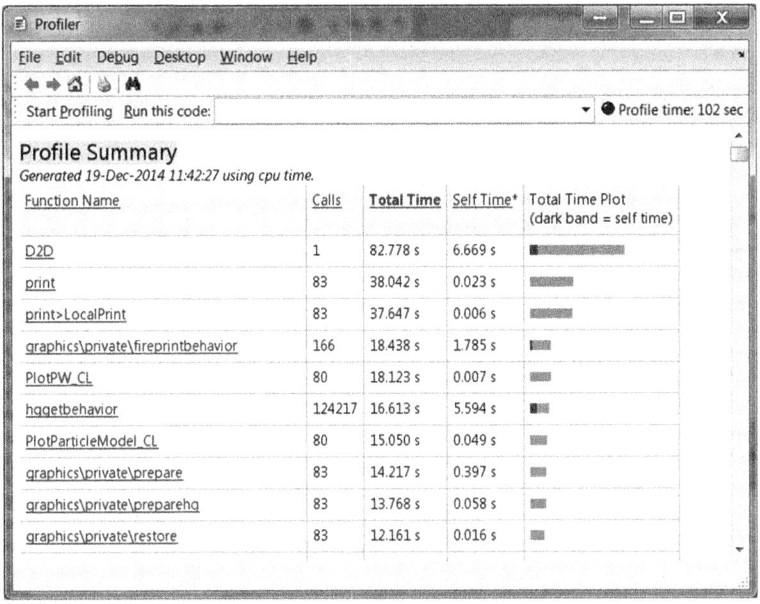

Profiler

File Edit Debug Desktop Window Help

Start Profiling Run this code: Profile time: 102 sec

Profile Summary

Generated 19-Dec-2014 11:42:27 using cpu time.

Function Name	Calls	Total Time	Self Time*	Total Time Plot (dark band = self time)
D2D	1	82.778 s	6.669 s	
print	83	38.042 s	0.023 s	
print>LocalPrint	83	37.647 s	0.006 s	
graphics\private\fireprintbehavior	166	18.438 s	1.785 s	
PlotPW_CL	80	18.123 s	0.007 s	
hggetbehavior	124217	16.613 s	5.594 s	
PlotParticleModel_CL	80	15.050 s	0.049 s	
graphics\private\prepare	83	14.217 s	0.397 s	
graphics\private\preparehg	83	13.768 s	0.058 s	
graphics\private\restore	83	12.161 s	0.016 s	

Figure 2.4. *Performance analysis of DICE2D (plotting and outputs enabled)*

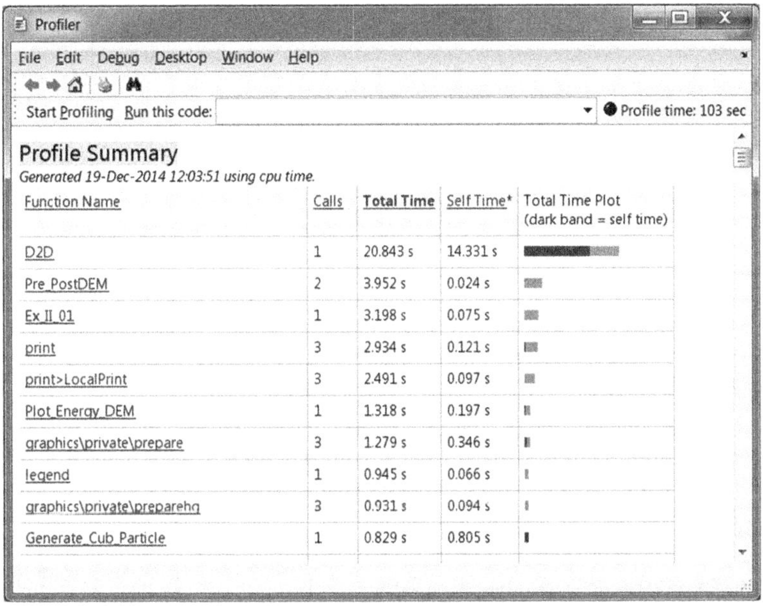

Profiler

File Edit Debug Desktop Window Help

Start Profiling Run this code: Profile time: 103 sec

Profile Summary

Generated 19-Dec-2014 12:03:51 using cpu time.

Function Name	Calls	Total Time	Self Time*	Total Time Plot (dark band = self time)
D2D	1	20.843 s	14.331 s	
Pre_PostDEM	2	3.952 s	0.024 s	
Ex_II_01	1	3.198 s	0.075 s	
print	3	2.934 s	0.121 s	
print>LocalPrint	3	2.491 s	0.097 s	
Plot_Energy_DEM	1	1.318 s	0.197 s	
graphics\private\prepare	3	1.279 s	0.346 s	
legend	1	0.945 s	0.066 s	
graphics\private\preparehg	3	0.931 s	0.094 s	
Generate_Cub_Particle	1	0.829 s	0.805 s	

Figure 2.5. *Performance analysis of DICE2D (plotting and outputs disabled)*

2.4. Parallel implementation

This section presents the parallel implementation of DICE2D based on *parfor*. The main purpose is to reduce the computational time of DEM simulation on multi-core PCs. As DEM is an explicit method in time, only minor changes are needed to parallelize the code. Quad-core PCs are quite common now, but the serial code cannot use its computing resources well. *Parfor* provides a useful tool to parallelize MATLAB code for a multi-core environment. A fork-join model is used in the parallelization. The work scheme of the serial and parallel DEM codes is shown in Figure 2.6. From the figure, it can be observed that the serial DEM has only one main thread. The contact detection and contact force calculations of the particles are performed sequentially (see Figure 2.6(a)). The multi-core DEM uses the fork-join model to let one cycle be calculated by more than one processor (abstracted as workers in MATLAB) (see Figure 2.6(b)). The parallel DICE2D works as follows. First, the client executes the preprocessing and activates all workers. Then, when the client requires parallel computing, the workers are allocated to the corresponding calculation tasks.

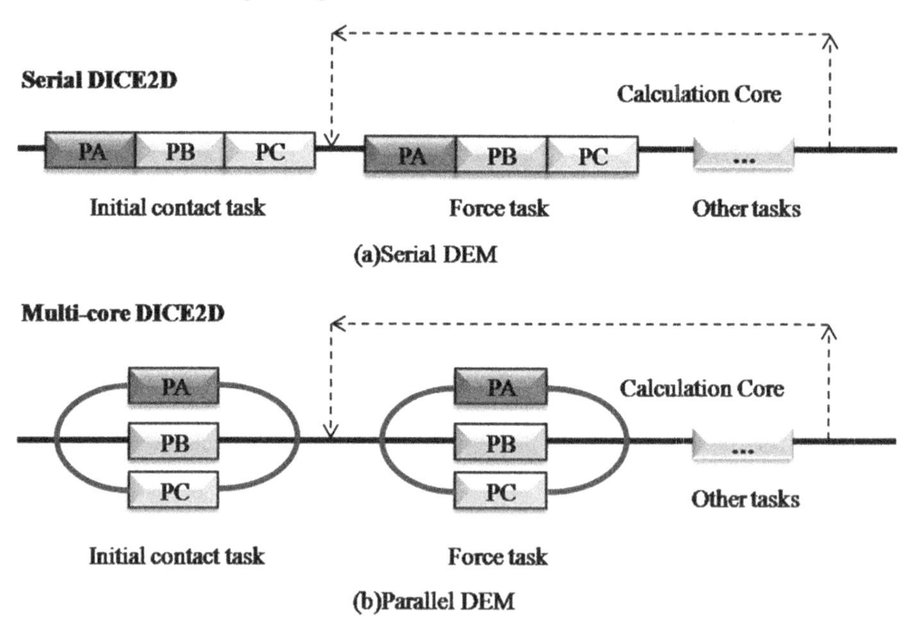

Figure 2.6. *Scheme of serial and parallel implementation of DEM in a multi-core PC*

From the profile analysis, it can be determined that the most time-consuming parts of DICE2D are the P2P contact force calculation and P2P contact detection. Therefore, the parallelization focuses on these two elements. For P2P contact force calculation, the code can be parallelized by simply replacing *for* with *parfor* (Table 2.4). Except for the keywords, the rest of the serial and parallel codes are the same.

Serial Code	Parallel Code
%Particle–particle contact force	%Particle–particle contact force
for j=1:NumP	**parfor** j=1:NumP
for k=1:MAX_PN	for k=1:MAX_PN
jP=PP_C(k,j);	jP=PP_C(k,j);
...	...
end	end
end	end

Table 2.4. *Parallelization of P2P contact force calculation*

However, this treatment may not work for other code segments. For example, error information was produced when the same approach was used for the P2W force calculation (Figure 2.7). This error occurred because MATLAB cannot recognize indirect address operations, for example when the calculation array is further indexed by another array. As the P2W calculation is not the main time-consuming component, parallelization is not used. The same problem is observed in the P2P contact detection. In the following sections, the solution to this problem is presented.

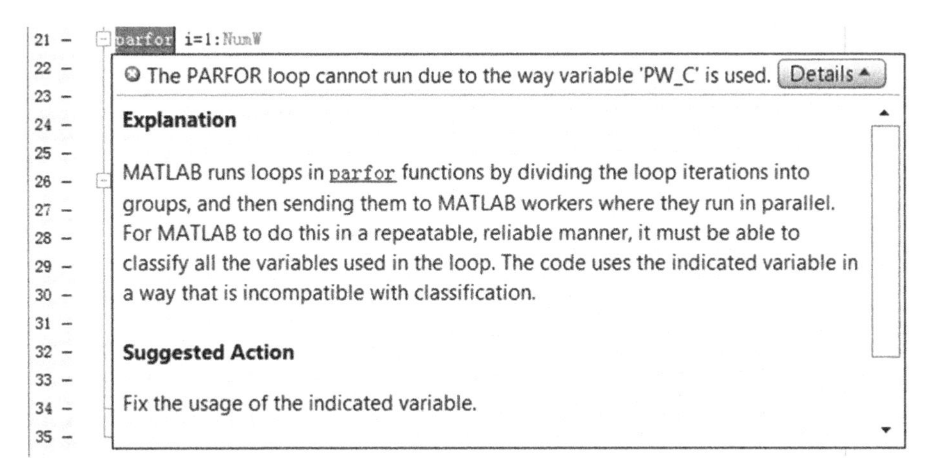

Figure 2.7. *Error information in parallelization of P2W force calculation*

For P2P contact detection, MATLAB cannot interpret the memory access method used in the contact list array (PP_C). Table 2.5 shows the original serial code of P2P contact detection and the corresponding parallel version. A temporary array is used to store the corresponding neighbor indexes. Instead of directly operating on the contact list, the code operates on the temporary array first and then passes the results to the contact list through a vector operation. Thus, the error information disappears, and the code can execute correctly.

Serial Code	Parallel Code
N=length(X);	N=length(X);
PP_C=zeros(MAX_PN,N);	PP_C=zeros(MAX_PN,N);
Index_Temp=ones(1,N);	Index_Temp=ones(1,N);
for i=1:N	**parfor** i=1:N
for j=1:N	Index_Temp=1;
if i==j	PLoc=zeros(MAX_PN,1);
else	for j=1:N
if abs(IDGX (i)–	if i==j
IDGX(j))<2 && abs(IDGY(i)–IDGY(j))<2	else
x1=X(i);	if abs(IDGX (i)–
y1=Y(i);	IDGX(j))<2 && abs(IDGY(i)–IDGY(j))<2
x2=X(j);	x1=X(i);
y2=Y(j);	y1=Y(i);
R1=R(i);	x2=X(j);
R2=R(j);	y2=Y(j);
if	R1=R(i);
Contact_P_P(x1,y1,x2,y2,R1,R2)>–dGap	R2=R(j);
	if
PP_C(Index_Temp(i),i)=j	Contact_P_P(x1,y1,x2,y2,R1,R2)>–dGap
	PLoc(Index_Temp)=j ;
Index_Temp(i)=Index_Temp(i)+1;	Index_Temp(i)=Index_Temp+1;
end	end
end	end
end	end
end	end
end	end

Table 2.5. *Parallelization of P2P contact detection*

The parallel implementation of DICE2D requires only a minor effort of replacing the corresponding keywords. The uniaxial compression test was used to determine the performance of the parallel DICE2D. The computational time and speedup of the parallel DICE2D are shown in Figure 2.8. When the particle number is less than

5,000 or more than 30,000, the performance of the parallel DICE2D is worse than that of the serial case because the computational time of the slice matrix operation and communication between the client and the workers is strongly influenced by the number of particles. When the number of particles is small, the computational gain from parallel computing is not comparable with the cost of the overhead time used for parallelization. However, when the number of particles is too large, the computational time spent on communication is again dominant due to the large array size; communication is slow and might be unable to fully use the high-speed cache of the computer. From Figure 2.8, it can be observed that models with approximately 10,000 particles can achieve the best performance of the parallel DICE2D. Overall, the performance of the multi-core (parallel) DICE2D implementation is better than that of the serial code.

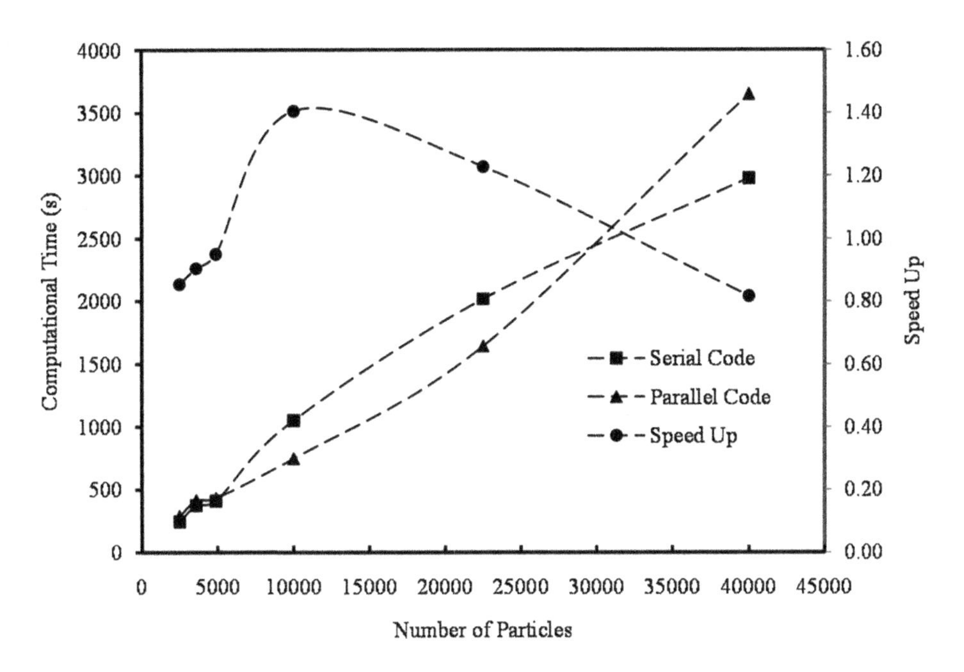

Figure 2.8. *Computational time and speedup analysis of multi-core (parallel) DICE2D*

Compared with the efforts put into parallelization, the parallel DICE2D is successful. However, nonlinear processes such as fracturing are not modeled in the numerical simulation. Moreover, the influence of the number of workers on the performance is not fully studied. In section 2.5, performance tests of the parallel DICE2D under different scenarios are presented.

2.5. Numerical examples

2.5.1. *Uniaxial compression test*

Despite the many advantages of DEM, micro parameter selection remains one of its main shortcomings [YOO 07]. In practice, a calibration process mimicking the actual physical test is adopted to determine the actual input micro parameters. In this example, Poisson's ratio will be determined using a numerical uniaxial compression test. The setup of the computational model is shown in Figure 2.9. A two-dimensional (2D) square specimen is represented by a group of circular particles. The corresponding physical configuration is a group of bonded cylindrical bars (with thickness of 1), which represents the plane stress condition. To simulate the plane strain problem, the following relationship can be used:

$$E' = \frac{E}{1-\nu^2}, \quad \nu' = \frac{\nu}{1-\nu} \qquad [2.1]$$

where E' and ν' are the corresponding elastic parameters under plane strain condition, and E and ν are the parameters reconstructed by the DEM.

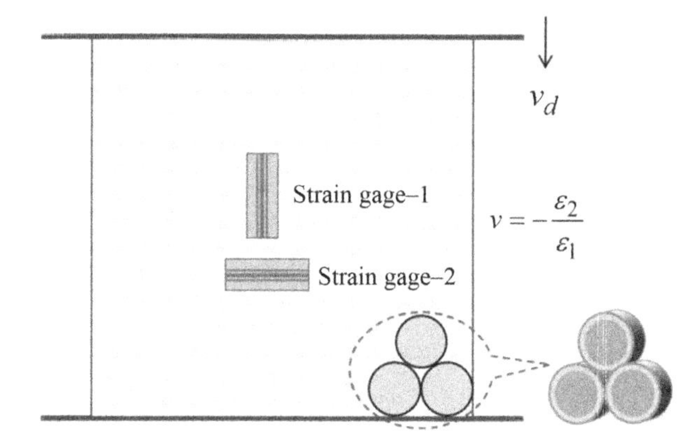

Figure 2.9. *Computational model to extract Poisson's ratio*

Displacements of two particles are recorded to represent the strain gage. The horizontal and vertical strains are calculated as follows:

$$\varepsilon_1 = \frac{u^A - u^B}{y_0^A - y_0^B}, \quad \varepsilon_2 = \frac{v^C - v^D}{x_0^C - x_0^D} \qquad [2.2]$$

where u^A and u^B are the y displacements of two measurement points, A and B; y_0^A and y_0^B are the initial y coordinates; and the definitions of v^C, v^D, x_0^C and x_0^D are analogous.

The first simulation is conducted on a regular-packed DEM (Figure 2.10(a)). A constant velocity loading (0.001 m/s) is applied to the top of the specimen, which will reproduce a strain rate of 10^{-5} s^{-1} (a quasi-static loading condition). Four points are selected to record the strain along the horizontal and vertical directions. The model parameters are listed in Table 2.6. The simulation results are shown in Figure 2.10(b)–(d). From the y displacement contour map of the model, as expected, a uniform distribution of the y displacement is obtained for the uniaxial compression. From the energy analysis, it can be found that the strain energy is the main energy variation. This result confirms the quasi-static loading scenario. Figure 2.10(d) shows the history of Poisson's ratio, which is always zero.

Number of particles	100	Normal viscous coefficient (s)	0
Mean particle size (m)	10	Local damping	0.8
Density (kg/m^3)	1,000	Wall tension	0
Normal stiffness (N/m)	1e8	Wall friction	0
Shear stiffness (N/m)	0	Wall cohesion	0
Tension strength (N)	5e9	Time step reduction factor	0.1
Cohesion	5e9	Total steps	800
Friction angle	80	Gravity acceleration (m/s^2)	0
Normal viscous coefficient (s)	0		

Table 2.6. *Model parameters of the uniaxial compression test of square-packed specimen*

Figure 2.11 shows the computational model and the results of a triangular-packed specimen. Except for the number of particles (95), the same model parameters listed in Table 2.6 are used. Unlike the square-packed specimen, Poisson's effect is observed (see Figure 2.11(d)). Therefore, Poisson's effect is strongly influenced by the packing pattern of the specimen. However, in the finite element method (FEM), the shape of the element does not influence Poisson's effect. For example, the rectangle element model can still reproduce the input Poisson's ratio. For further investigation, a random-packing specimen is simulated, and the results are shown in Figure 2.12. Poisson's effect is also reproduced by the irregular-packed specimen (see Figure 2.12(a)). To obtain Poisson's ratio, the value of the stable range of the history curve is adopted (see Figure 2.12(d)).

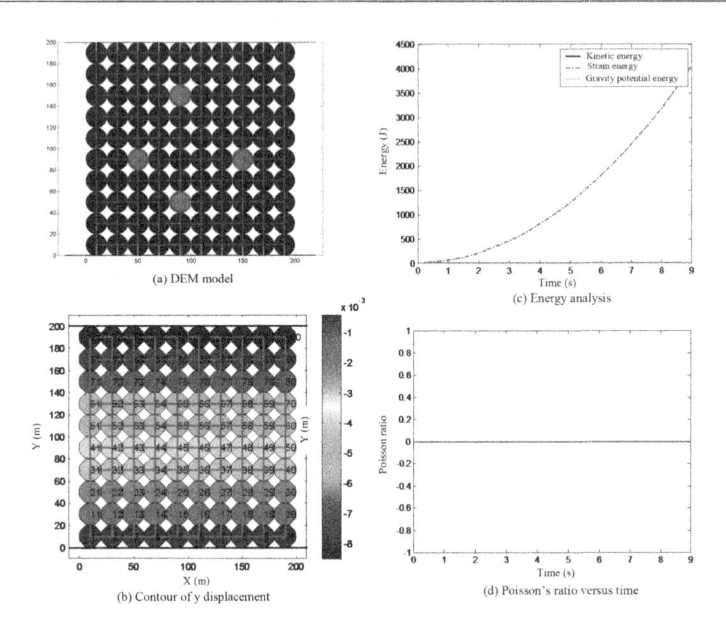

Figure 2.10. *Computational model and simulation results on a regular-packed DEM sample. For a color version of the figure, see www.iste.co.uk/zhao/computing.zip*

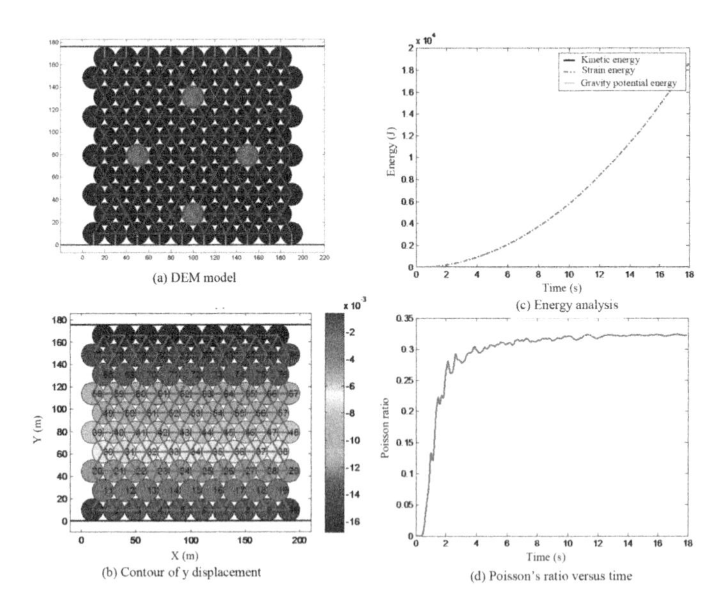

Figure 2.11. *Computational model and simulation results on a triangle-packed DEM sample. For a color version of the figure, see www.iste.co.uk/zhao/computing.zip*

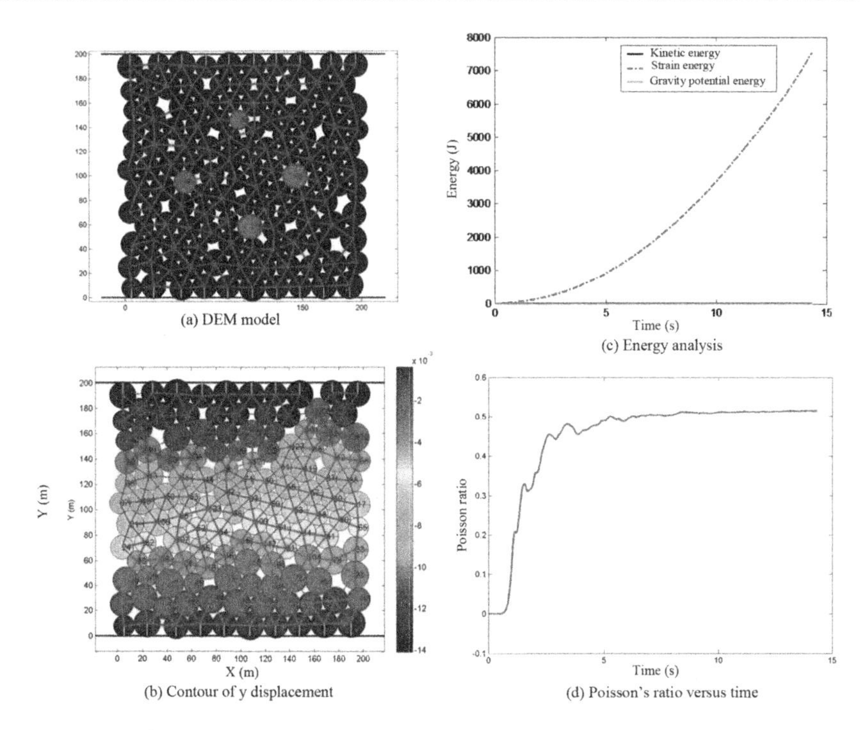

(a) DEM model

(c) Energy analysis

(b) Contour of y displacement

(d) Poisson's ratio versus time

Figure 2.12. *Computational model and simulation results on an irregular-packed DEM sample. For a color version of the figure, see www.iste.co.uk/zhao/computing.zip*

From these DEM simulations, Poisson's ratio effect is a result of the force chain. For an irregular-packed specimen, the coordinate number (CN) (average number of bonds of each particle) is used to characterize the structure difference between models with different threshold values for bond forming. From previous studies (e.g. [YOO 07]), Poisson's ratio is found to be directly controlled by the ratio between shear and normal stiffness, K_s/K_n. Figure 2.13 shows the relationship between K_s/K_n and Poisson's ratio of different particle packing models. For the regular square-packed specimen, Poisson's ratio is always zero. For the regular triangular-packed specimen, Poisson's ratio can be adjusted using different K_s/K_n. It should be noted that there is a limit of 1/3. Poisson's ratio for irregular packing can be larger than this limit. However, a particle model with large CN will produce a specimen similar to the regular triangle-packed specimen (see Figure 2.13). Therefore, for a densely packed specimen (CN is large), there should be a Poisson's ratio limitation for DEM, which is similar to the classical lattice spring model (LSM) proposed by Hrennikoff [HRE 41]. Therefore, the disadvantage of Poisson's limitation in LSM is still not completely overcome in DEM.

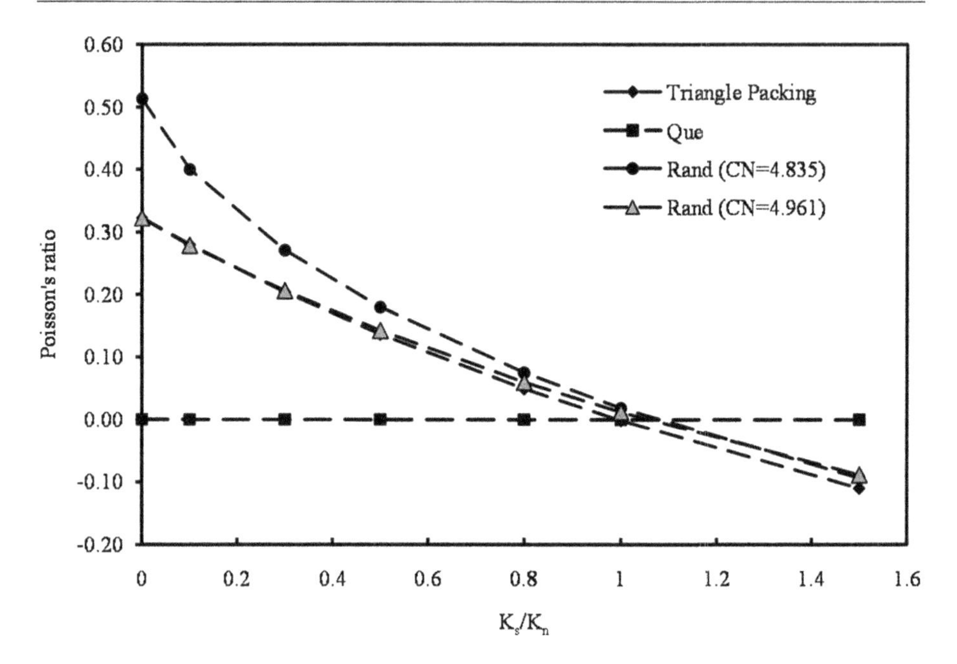

Figure 2.13. *Influence of K_s/K_n on Poisson's ratio of different particle models*

2.5.2. *Beam bending test*

In this example, the elastic modulus of DEM is studied. The commonly used uniaxial compression test in the previous example is a one-dimensional (1D) controlled problem that does not consider the influence of rigid body rotation. For example, the computational model in Figure 2.10(a) cannot actually represent the continuum model with a Poisson's ratio of zero because the square-packed model is unstable under bending loading. The beam bending problem (Figure 2.14) involves shear, compression and tension states, and is a good candidate for the calibration of elastic modulus.

The analytical solution of the dimensionless deflection is given as follows:

$$\begin{cases} \bar{u}_y(a) = \left(a\left(4a^2 - 3\right)\right), & a < 0.5 \\ \bar{u}_y(a) = (1-a)\left(4(1-a)^2 - 3\right), & \text{else} \end{cases} \qquad [2.3]$$

where $\bar{u}_y(a) = u_y(a) / \left|u_y^{\max}\right|$ is the dimensionless beam deflection along the loading direction, $u_y(a)$ is the deflection, $u_y^{\max}$ is the maximum deflection, $a = x / L$ is the dimensionless x coordinate and L is the support length. The elastic modulus can be obtained by:

$$E = \frac{FL^3}{48u_y^{\max} I_{\text{BEAM}}} \qquad [2.4]$$

where F is the loading force and I_{BEAM} is the inertial cross area that can be calculated as follows:

$$I_{\text{BEAM}} = \frac{H^3}{12} \qquad [2.5]$$

where H is the height of the beam.

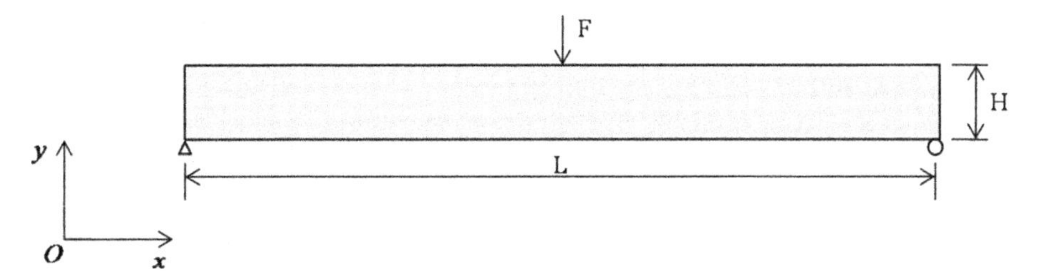

Figure 2.14. *The beam bending problem*

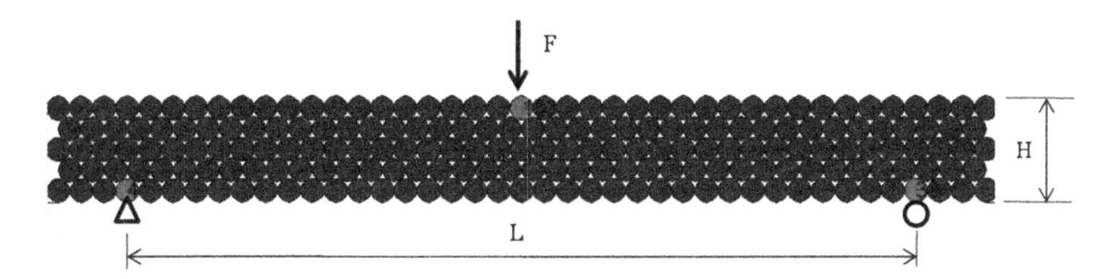

Figure 2.15. *DEM computational model of the beam bending problem*

The main purpose of the example is to study the elastic moduli of DEM models and to verify the ability of DICE2D to model elasticity problems. It should be noted that equation [2.3] is only valid under small deformation. Therefore, the applied force is controlled at a relatively small value. Figure 2.15 shows the computational model to simulate the beam bending problem. From the previous example, it is clear that the square-packed model is not suitable for continuum modeling, whereas the triangular-packed model provides a good reference for continuum modeling and parameter selection (see Figure 2.13). In this example, the beam is constructed based on the triangular-packed particle model. Roller and fixed boundaries are applied to the beam bottom through particle boundary conditions applied to two particles. The loading force is applied to one particle on the top, as shown in Figure 2.15. A number of measurement points are selected along the middle line of the beam to record the deflection during calculation.

It should be noted that, for a triangular-packed model, when shear spring and rigid body rotation are not considered, an analytical solution between the elastic modulus and the normal stiffness can be written as follows [HRE 41]:

$$E = k_n \left(\frac{1}{\Delta\sqrt{3}} \right) \qquad [2.6]$$

where Δ is the unit thickness required to maintain dimensional consistency. The model parameters of the beam simulation are shown in Table 2.7. It should be noted that the shear stiffness will assume different values when investigating the influence of K_s/K_n on the elastic modulus. The bond thickness ratio is another parameter that must be investigated. Using equation [2.6], the estimated elastic modulus is found to be 57 MPa.

Number of particles	203	Cohesion	13e9
Mean particle size (m)	10	Friction angle	80
Density (kg/m^3)	1,000	Local damping	0.8
Normal stiffness (N/m)	le8	Time step reduction factor	0.1
Shear stiffness (N/m)	–	Total steps	800
Tension strength (N)	13e9	Gravity acceleration (m/s^2)	0

Table 2.7. *Model parameters of the beam bending problem*

Figure 2.16 shows the energy analysis and simulation results of the beam bending problem when $K_s/K_n = 0.2$ and WRT = 0.2. The strain energy of the beam reaches a relatively stable state, whereas the kinetic energy approaches zero at the end of the simulation. These conditions ensure that the modeling is in a quasi-static condition. Using equation [2.4], the elastic modulus is calculated to be 40 MPa. The difference is due to the actual depth of the beam being less than the value used in Figure 2.15, which is a geometric property of the particle model. The particle force is applied at the center of the loading particle rather than at its surface. In DEM, the difference between these two cases cannot be distinguished. The DEM prediction and analytical solution of the dimensionless deflection are shown in Figure 2.16(b). Good agreement is achieved. It can be concluded that the DEM can reproduce a reasonable elastic solution.

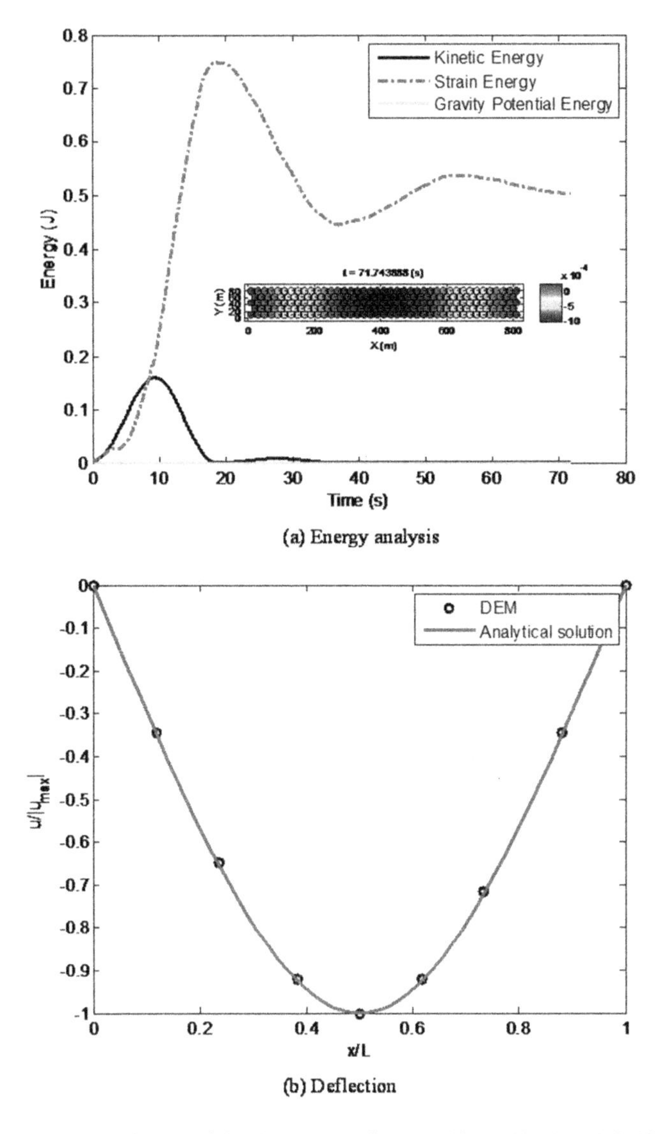

(a) Energy analysis

(b) Deflection

Figure 2.16. *DEM modeling of the beam bending problem (K_s/K_n = 0.2, WRT = 0.2). For a color version of the figure, see www.iste.co.uk/zhao/computing.zip*

From the uniaxial compression test simulation, it was found that WRT does not influence Poisson's ratio. However, numerical results from the beam bending example show that WRT influences the elastic modulus. Nevertheless, this indication may not be as true as the data suggest for the following reason. Due to the influence of the geometric characteristic of the particle model, the effective beam height is around *H-D* when WRT is zero (*D* is the particle diameter). When WRT is larger, the effective beam height will approach *H*. According to equations [2.4] and [2.5], a greater elastic modulus will be obtained for a larger WRT. This variation decreases when the number of particles increases (higher resolution).

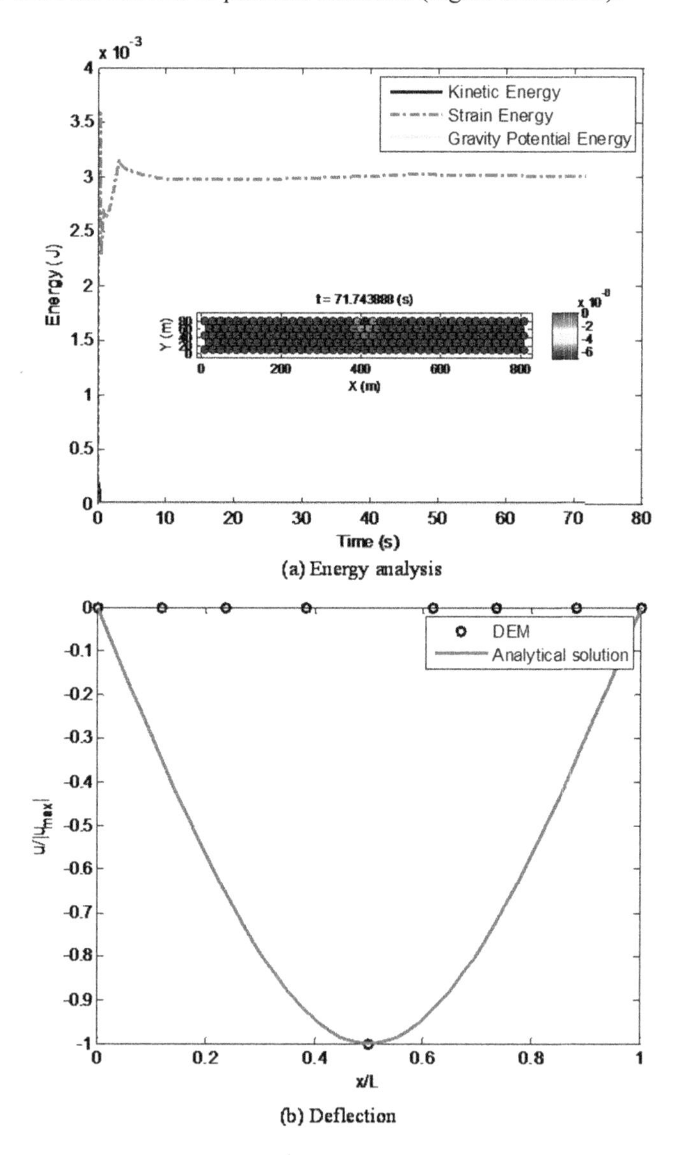

Figure 2.17. *DEM modeling of the beam bending problem (K_s/K_n = 0.2, WRT = 1.0). For a color version of the figure, see www.iste.co.uk/zhao/computing.zip*

Figure 2.17 shows the energy analysis and simulation results of the beam bending problem when WRT is 1.0. From the strain energy and kinematic energy, it can be observed that a stable convergent solution is obtained. However, from the contour map of the *y* displacement (see Figure 2.17(a)), it can be found that the distribution is not the classical distribution of the beam bending problem (see Figure 2.16(a)). The predicted dimensionless deflection and analytical solution are shown together in Figure 2.17(b). From these results, it can be concluded that the DEM with WRT of 1.0 cannot reproduce an elastic solution. In other words, the elasticity is not applicable to the discrete model. From a simple parameter study, it is found that an elastic solution can be obtained when WRT is within 0.0–0.9. In this book, 0.1 is suggested as the default value for DEM simulations of cohesive granular materials such as rock and concrete.

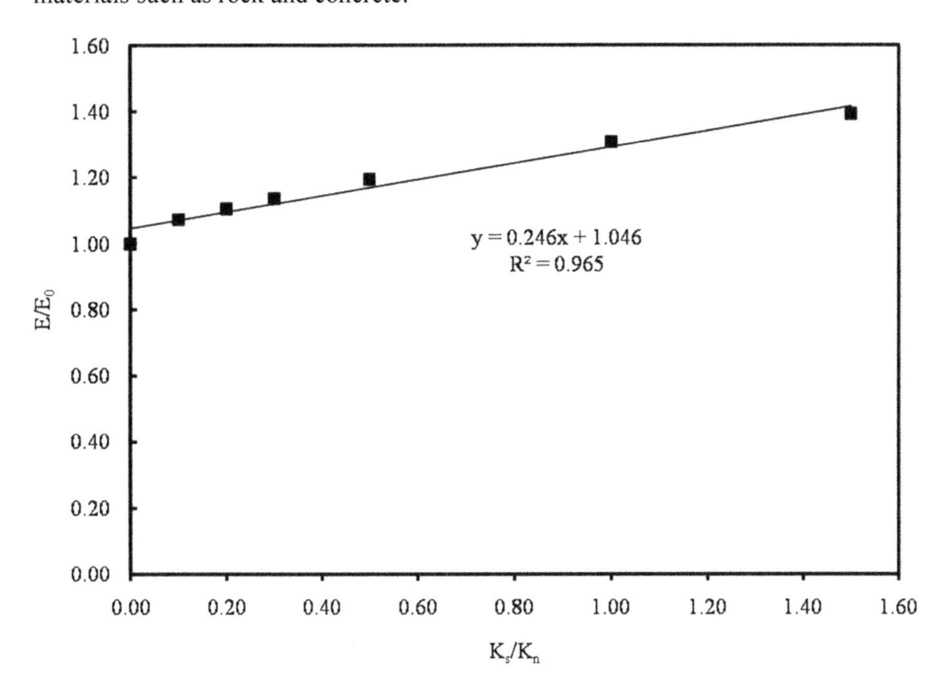

Figure 2.18. *Relationship between K_s/K_n and the elastic modulus calculated from the beam bending problem*

From the earlier analysis, K_s/K_n is selected as the control variable to link with the elastic modulus. When K_s/K_n is zero, the elastic modulus of the model is marked as E_0. Then, all elastic moduli of other K_s/K_n values are scaled as E/E_0. The relationship between K_s/K_n and the scaled elastic modulus is shown in Figure 2.18. A simple linear function can be found. In actual modeling, the following procedure can be used to estimate the micro mechanical parameters in DICE2D. First, the value for K_s/K_n can be estimated from Figure 2.13 using a graphic method with the target Poisson's ratio. Then, E/E_0 can be further obtained from Figure 2.18. Finally, K_n can be obtained from equation [2.5]. A good estimation of K_n and K_s can make the calibration process faster. Moreover, for problems in which deformation is not the main concern, these estimated values might be directly used in actual modeling.

The beam bending problem is executed five times using the parallel DICE2D with a different number of workers. When the number of workers is zero, the parallel DICE2D works as a serial DEM code. The computational times are listed in Table 2.8. When the number of workers increases, the computational time generally decreases. However, 3 seems to be an optimal number, which is because the laptop has four CPUs. If the client uses one, only three cores are left for the workers. In this example, the serial code runs much faster than the parallel code. Because a small number of particles are used in this example, the benefit obtained from parallel computing on calculation time saving does not exceed the overhead time spent on parallelization.

Number of workers	0	1	2	3	4
Computational time	194.87	2554.58	1042.61	714.54	781.35

Table 2.8. *Computational time of the beam bending problem by parallel DICE2D using different number of workers*

For both the uniaxial compression test and the beam bending problem, the contact detection during calculation is not active. The performance benchmark in Figure 2.8 and Table 2.8 is true only for elastic intact problems. In section 2.5.3, a problem involving dynamic contact detection is solved to test the performance of the parallel DICE2D on modeling granular flow-like problems.

2.5.3. *Collapse of a granular tree under gravity*

In this example, a case involving massive contact detection during calculation is shown. As I prepared this example during Christmas time, the model was made up as a Christmas tree. The image-based modeling technique was adopted. The basic principle is to build up base particles. Then, a subroutine is used to filter out the background particles. The MATLAB implementation is shown in Table 2.9. Figure 2.19 shows the digital image of a Christmas tree. The corresponding DEM model built from the image model subroutine is shown in Figure 2.20. Three walls are used to make up a container for the collapsed particles.

```
function [X,Y,R,T,MAT]=Generate_Tree_Particle
% Build a particle model form image
A=imread('Tree.bmp');
[n,m]=size(A);
Nnum=sum(sum(1-A));
iIndex=1;
X=zeros(1,Nnum);
Y=zeros(1,Nnum);
R=10*ones(1,Nnum);
T=zeros(1,Nnum);
MAT=ones(1,Nnum);
for i=1:n
for j=1:m
if(A(i,j)==0)%Not background (white refers background)
        xCur=10+(j-1)*20;
        yCur=20*n-(10+(i-1)*20;
        X(iIndex)=xCur;
        Y(iIndex)=yCur;
        iIndex=iIndex+1;
end
end
end
```

Table 2.9. *MATLAB implementation of building a DEM particle model from an image*

Figure 2.19. *Digital image of a Christmas tree*

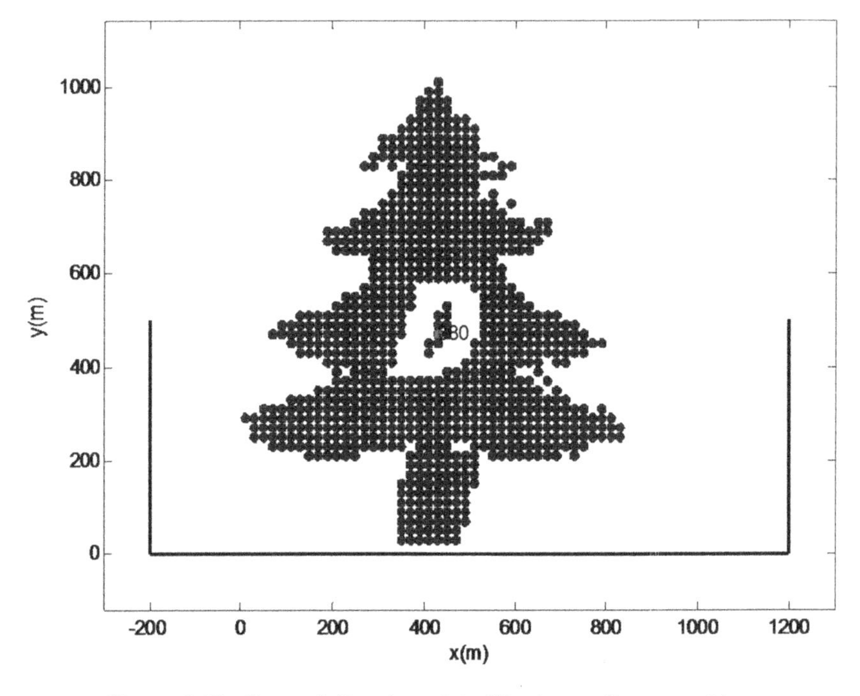

Figure 2.20. *Computational model of the tree collapse problem*

The model parameters are shown in Table 2.10. The strength parameters are set to be zero. Under gravity, the tree breaks down and turns into a granular-like flow. Figure 2.21 shows the tree collapse process. The tree initially collapses under gravity (see the first row images). Then, the upper part expands into a much looser granular state, which is further compacted under gravity. The rest of the process looks like a granular flow. This process can also be used as a particle packing method. In section 2.5.4, the final particle model of Figure 2.21 is used as the initial particle model for the next example.

Number of particles	846	Normal viscous coefficient (s)	0.001
Mean particle size (m)	10	Friction angle	0
Density (kg/m^3)	1,000	Local damping	0.01
Normal stiffness (N/m)	1e8	Time step reduction factor	0.1
Shear stiffness (N/m)	1e7	Total steps	20,000
Tension strength (N)	0	Gravity acceleration (m/s^2)	10
Cohesion	0		

Table 2.10. *Model parameters of the tree collapse problem*

This example was also simulated by the parallel DICE2D using a different number of workers. The computational times are shown in Figure 2.22. A speedup factor of approximately 2 is achieved when the number of workers is 3, because contact detection is the most time-consuming part of this example. It can be concluded that the contact detection of DICE2D has been successfully improved in the parallel DICE2D.

2.5.4. *Block caving*

Block caving is considered to be the most economic choice for large-scale underground mining, which costs only about 10% (~5–10 AUDs per ton) of the cost of stopping methods (~30–60 AUDs per ton). This approach involves a number of classical rock mechanics problems, for instance deformation of rocks under excavation at great depth, fracturing of rock under dynamic and quasi-dynamic loading, fragmentation of fractured rock under gravitational force and granular flow of rock fragments under gravity. These in turn control many crucial aspects relating to block caving, such as stability and serviceability of undercuts and draw horizons; caveability and production; and ground surface subsidence. The ground surface subsidence can further trigger additional serious geotechnical hazards and may jeopardize mine infrastructure. One example was reported in the Palabora copper mine, where a 300 m landslide was trigged by a block caving operation, affecting the water and power lines, a railway line and water reservoirs. Three approaches are available to analyze rock mechanics problems in block caving: empirical methods, experimental methods and numerical methods. Using the design charts with the design parameters (e.g. mining rock mass rating, height of the caved rock, and minimum and maximum spans of the footprint), the caveability, production and ground surface subsidence of block caving can be approximated in a simple manner. The main shortcoming of the empirical approach is the difficulty of determining the parameters related to rock masses, for example the mining rock mass rating and the density of fractured rock. Furthermore, the empirical method also ignores the stress–strain relationship of the rock masses and the influence of geological structures, and other site-specific issues that can affect the actual caving behavior significantly.

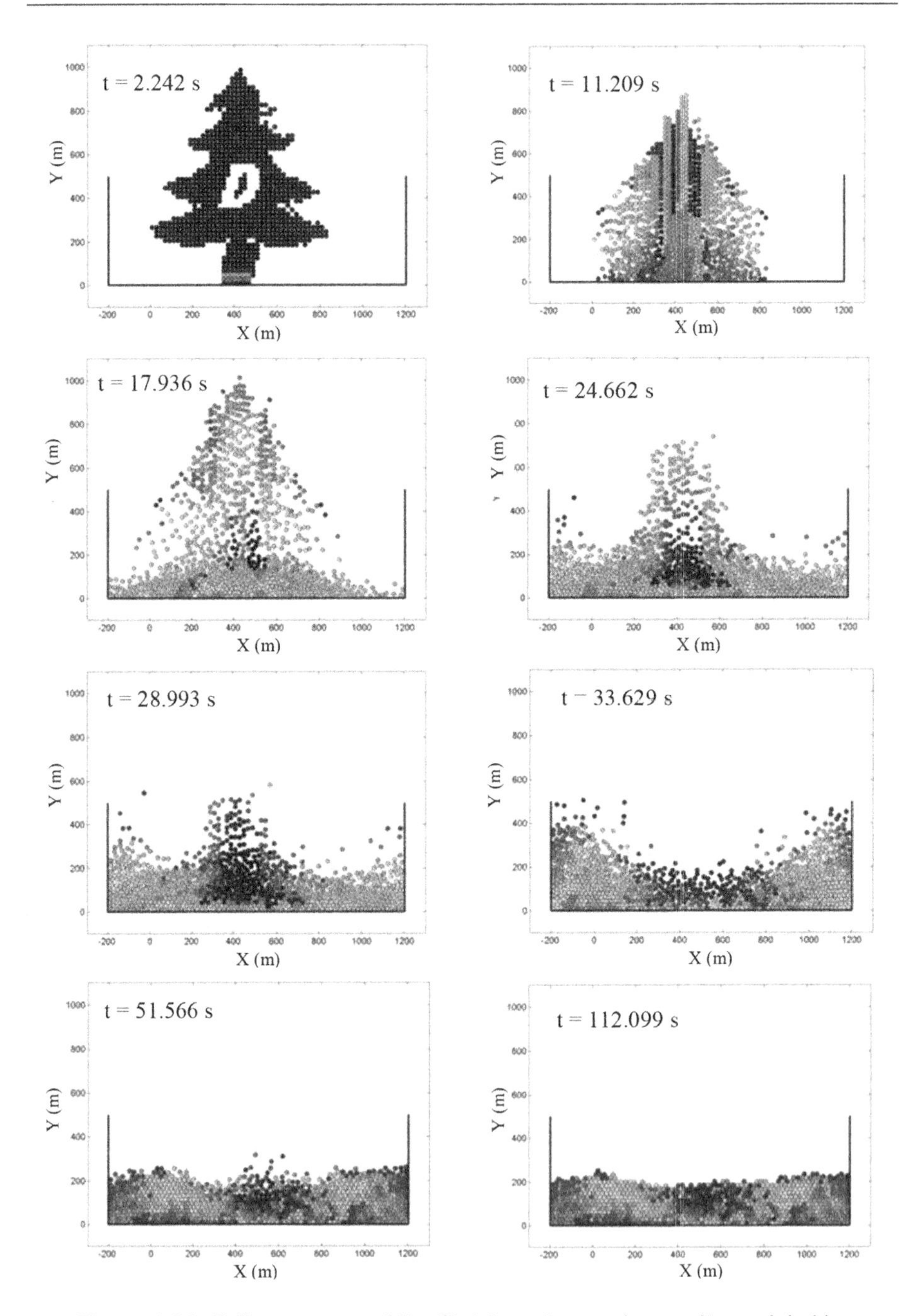

Figure 2.21. *Falling process of the Christmas tree under gravity modeled by DICE2D. For a color version of the figure, see www.iste.co.uk/zhao/computing.zip*

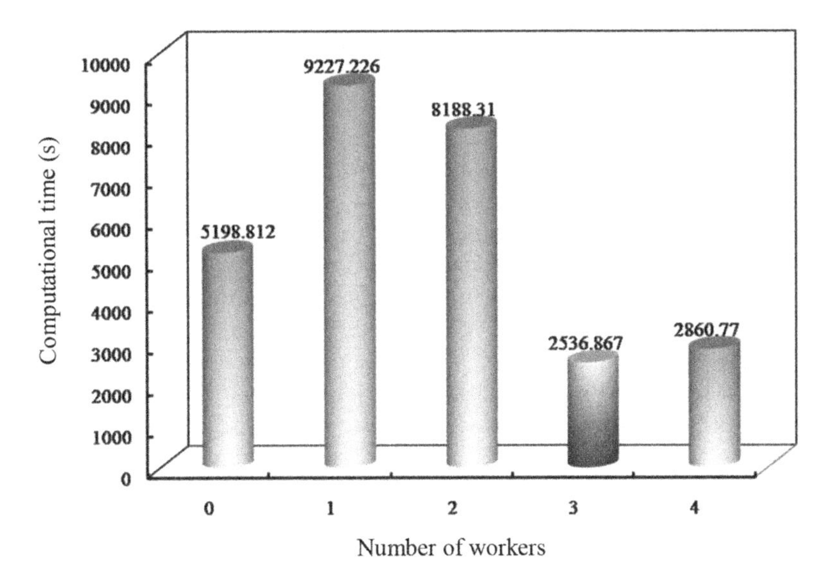

Figure 2.22. *Computational time of the tree collapse problem using parallel DICE2D with different number of workers*

The experimental approach can provide physical insights into the behavior of rock masses mined in block cavings. However, due to the large expense and time in the model construction, only a few tests on block cavings can be found in the literature, for example the 2D caving model tests conducted by McNearny and Abel [MCN 93] from Colorado School of Mines, and the three-dimensional (3D) model tests conducted by Trueman *et al.* [TRU 08] from the University of Queensland. These physical models provided useful information on the response of rock masses during block caving (e.g. the failure patterns of the caving zone and the deflection of the whole model including the ground surface subsidence). However, given the cost associated with each test and the construction time, physical tests are cost prohibitive for practical purposes. Moreover, some unrealistic assumptions must be made in the model testing, for example the horizontal stresses are not simulated correctly, the blocks are arranged uniformly [MCN 93] and the rock mass is in a discrete/granulated state without undergoing failure or fracturing [TRU 08]. These assumptions lead to model test results that can only be used for research purposes rather than a predictive model to guide the actual operation in block caving.

With the improvement of modern computers and computing power, numerical modeling techniques have become exceptionally useful in scientific research and engineering applications, and provide the most promising solution to study mechanical behavior of rock masses [BRO 08]. However, there are many limitations in current numerical techniques and, in practice, empirical methods, such as

Laubscher's method [LAU 00], are still the most commonly used methods in block caving. For example, the FEM, as the mainstream numerical tool in scientific research and engineering applications, is still limited in modeling fracturing and fragmentation of rock masses due the lack of sophisticated constitutive models for rock mass and the difficulty in parameter selection. The DEM is promising in simulating the complex mechanical interactions of rock masses such as fracturing and fragmentation. Nevertheless, a major shortcoming of the DEM is that proper calibration of the model parameters is required to obtain reasonable results [CAM 13]. In addition, due to the lack of advanced constitutive models for the DEM, it is unlikely that the DEM with a large element size (required for practical problems) can capture the nonlinear deformation of rock masses at the pre- and post-failure stage. The FEM/DEM [MUN 95] is a newly developed method to integrate FEM and DEM while avoiding their disadvantages. However, implementing this method into a computer code requires complex routines. Moreover, there are 12 DOFs for each numerical unit of the 3D FEM/DEM (6 DOFs for DEM) and is computationally costly. In addition, proper calibration is still required for the FEM/DEM to model fracturing and fragmentation; furthermore, a sophisticated constitutive model is still needed for the FEM/DEM to realistically model the nonlinear deformation of rock masses.

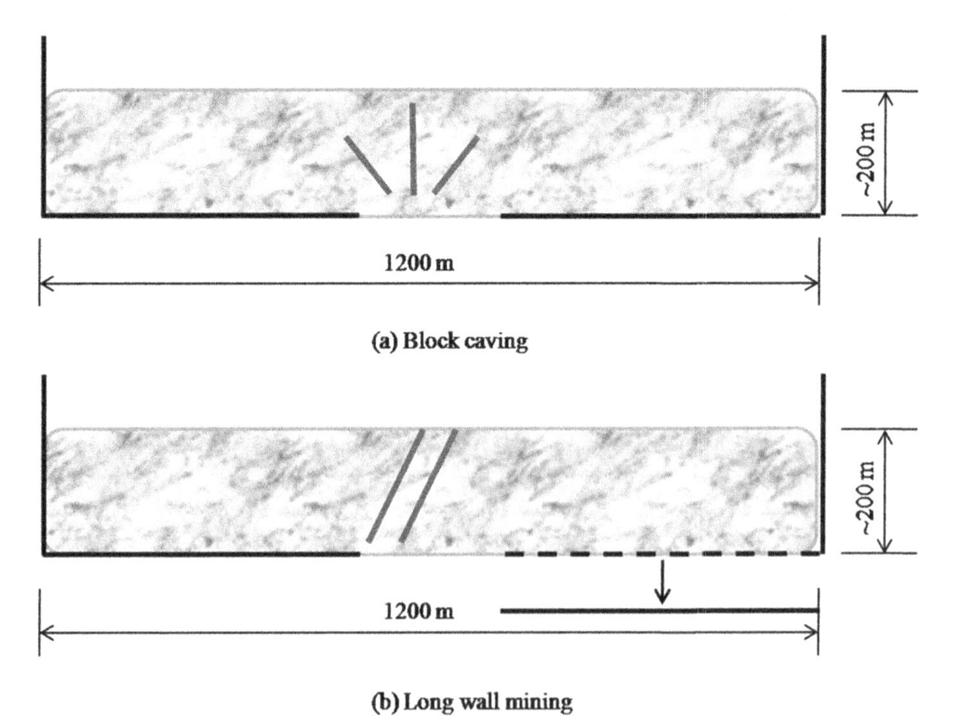

Figure 2.23. *Simplified model configurations for block caving and long wall mining*

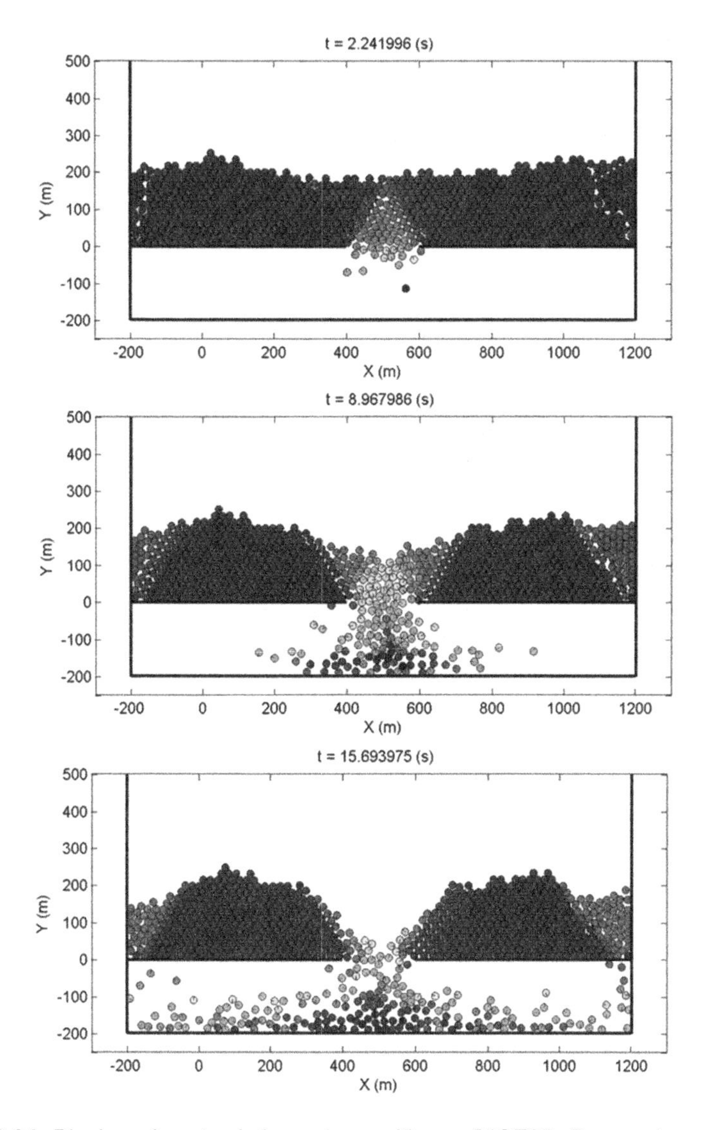

Figure 2.24. *Block caving simulation using multi-core DICE2D. For a color version of the figure, see www.iste.co.uk/zhao/computing.zip*

In this example, DICE2D is preliminarily used to model the block caving process. The computational model of the block caving model used in DICE2D is shown in Figure 2.23(a). A long-wall mining model is built as shown in Figure 2.23(b) to provide a comparison. The particle model uses the final packed particles in the previous example. In the block caving, a portion of the middle wall is removed to further fracture the ore body. In contrast, in the long-wall mining simulation, the right support wall is moved downward during the calculation. Figures 2.24 and 2.25 show the failure process of the rock masses under these two conditions. It can be found that the failure of the block mining is a granular-like type. The long-wall mining method first makes fractures in the continuum and then breaks it into blocks that will further be broken into small pieces.

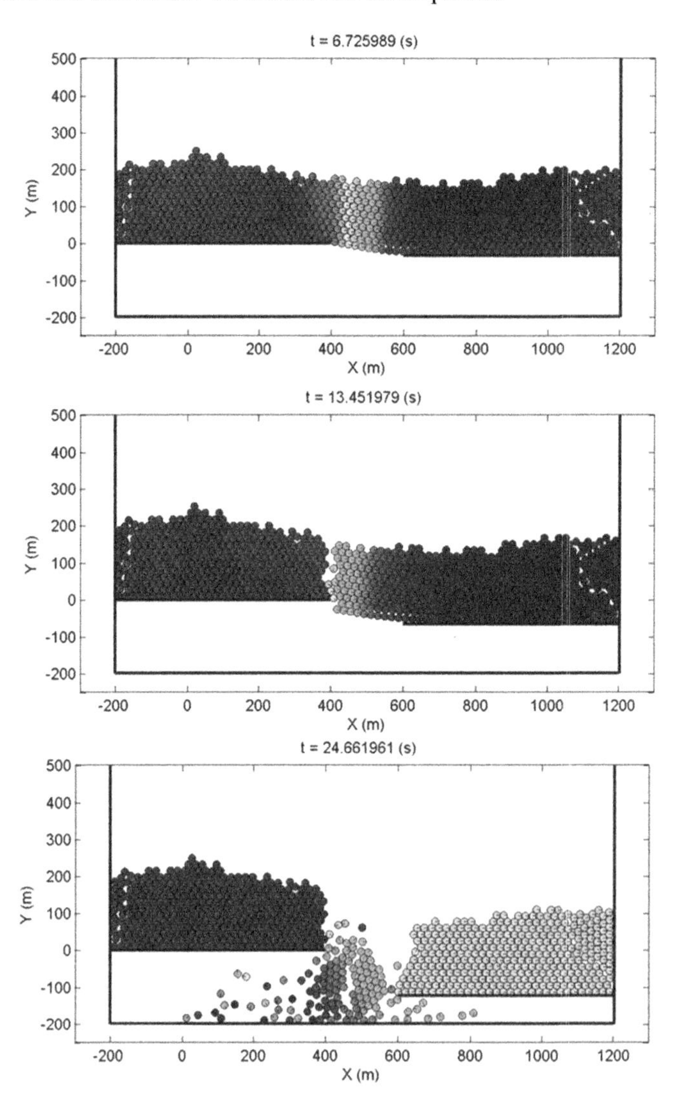

Figure 2.25. *Long-wall mining simulated by multi-core DICE2D. For a color version of the figure, see www.iste.co.uk/zhao/computing.zip*

2.6. Conclusion

In this chapter, DICE2D is parallelized for a multi-core PC. The Parallel Computing Toolbox of MATLAB is used as the development tool. Both the contact detection and the contact force calculation modules of DICE2D are parallelized. Numerical examples show that the parallel (multi-core) DICE2D can reduce the computational time effectively. A maximum speedup of approximately 2 can be achieved, which is successful compared with the efforts put into the parallel implementation. The numerical examples also verified the abilities of DICE2D on modeling continuum deformation, granular flow and fragmentation problems. Moreover, an empirical procedure is also presented to estimate spring stiffness from the corresponding macro parameters (Poisson's ratio and the elastic modulus).

3

GPU Implementation

In this chapter, DICE2D is implemented with the GPU of a computer. First, the principle of GPU computing is introduced briefly. Then, the GPU implementation of DICE2D is described. Finally, numerical examples are presented to show the performance improvement and application of the GPU DICE2D.

3.1. Graphics processing unit computing

Currently, computing on a GPU chip is popular for numerical modeling in different areas (e.g. elasticity simulation [DIC 11], fluid mechanics [HOR 11] and black hole simulation [HER 11]). An attractive aspect of GPU computing is the capability of parallel computing without the complex implementation of model decomposition, communication and synchronization functions. A GPU uses multiple threads rather than multiple CPUs for computing; this method is also called fine-grained parallelization. Compared with a multi-core CPU, a GPU card consists of a large number of low-level processors; for example there are 96 cores on a Quadro 600 graphics card, 512 cores on the GeForce GTX 580 graphics card and 5,760 cores on the GTX TITAN Z graphics card (www.nvidia.com). For each GPU processor, there are several hundreds of co-resident threads that can execute integer-, single- and double-precision calculations simultaneously. In addition, the memory access method (e.g. memory coalescing) is specially designed in GPUs to improve memory access performance.

GPU parallelization of various numerical methods has been implemented by many researchers, including the molecular dynamics [AND 08, STO 10], the lattice Boltzmann method [WAL 09, SAI 10], the FEM [JOL 10, KOM 10, DIC 11], the boundary element method [TAK 09], the DEM [MA 11, XU 11], upper bound rigid block analysis [POD 11], the moving particle semi-implicit method [HOR 11], parallel drainage network computation [ORT 10] and the distinct lattice spring model (DLSM) [ZHA 12]. The overall increases in speed using GPU parallelized codes compared to the serial CPU counterparts are reported to be between 10- and 100-fold. However, a substantial amount of knowledge regarding GPU hardware architecture is required to directly implement a GPU code. To overcome this problem, several types of development tools have been designed (e.g. special programming toolkits listed by Elsen *et al.* [ELS 08], including Sh (Michael McCool, University of Waterloo), Brook (Pat Hanrahan, Stanford University), Close to Metal (AMD) and CUDA (NVIDIA)). Among these toolkits, CUDA is the most popular; however, it is not directly supported in MATLAB®. In this chapter, to reap the benefit of GPU computing, DICE2D is parallelized using the Parallel Computing Toolbox® of MATLAB.

GPU computing is based on the heterogeneous computing methodology. The primary code runs on the CPU (i.e. the host), while the computationally expensive components of the code run on the GPU (i.e. the device). The hardware architecture of a GPU computer is shown in Figure 3.1. As shown in the figure, the device can be viewed as a virtual computer that has its own separate memory space and is ideally suited to perform arithmetic calculations using thousands or millions of elements. The GPU and CPU are connected through a peripheral component interconnect (PCI). In GPU implementation, there are two essential concerns: the first is how the data between the memory of the CPU and that of the GPU should be communicated; and the second is how the GPU should be instructed to perform the corresponding calculations. In MATLAB, these implementations have been significantly simplified; the Parallel Computing Toolbox of MATLAB provides GPU matrix computing and communication functionalities. Basic calculations using a GPU matrix are supported in MATLAB, which are in the same form as the corresponding CPU versions. More details on GPU matrix computing can be found in the manual of the Parallel Computing Toolbox of MATLAB.

3.2. GPU implementation of DICE2D

The GPU implementation of DICE2D is to replace corresponding matrix calculations with GPU matrix calculations. The first step is to define the GPU matrices. In MATLAB, there are two methods to define a GPU matrix: the first is to define a CPU matrix initially, and then send it to the GPU and duplicate it. For example, the code shown in Table 3.1 can be used to build a 2×2 GPU matrix with zero elements.

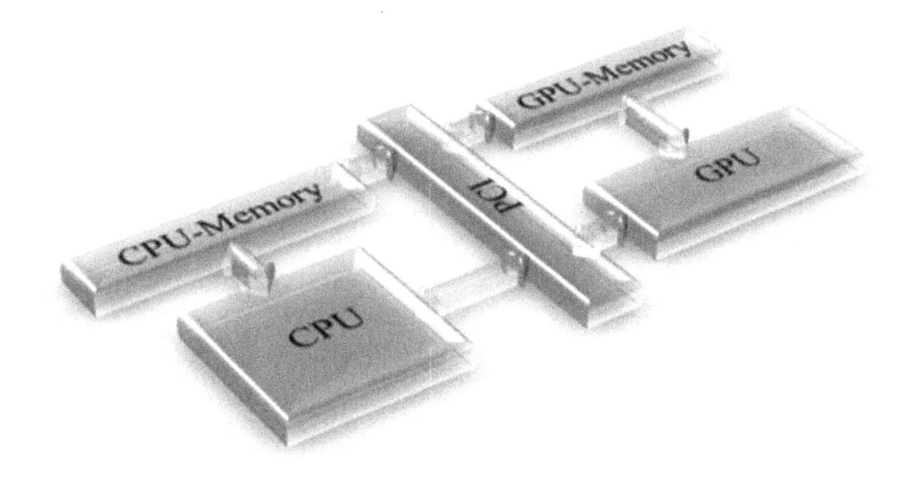

Figure 3.1. *Hardware configuration of the GPU computing*

```
cpuMat=zeros(2,2)

gpuMat=gpuArray(cpuMat);
```

Table 3.1. *Define a GPU matrix in MATLAB (Method 1)*

The second method is to directly define a GPU matrix using the built-in GPU function. One example code is shown in Table 3.2.

```
gpuMat=parallel.gpu.GPUArray.zeros(2,2);
```

Table 3.2. *Define a GPU matrix in MATLAB (Method 2)*

To receive data from the GPU, the code shown in Table 3.3 can be used.

```
gpuMat=gather (gpuMat);
```

Table 3.3. *Read matrix from GPU to CPU in MATLAB*

To show the performance improvement of the GPU computation, two example MATLAB codes (one serial code and the corresponding GPU code) are programmed (Table 3.4). Unlike in C++ [ZHA 12], the parallel computing settings (e.g. number of blocks and number of threads per block) are done by MATLAB automatically. The increased speed of the GPU code is shown in Figure 3.2. It is shown that the GPU code obtained a maximum speed increase of approximately sixfold. When the array size is too small or too large, the performance of the GPU code tends to decrease, which might be because the communication between the CPU and the GPU requires more computation time than the benefit that the GPU computation provides under these conditions.

CPU code	GPU code
function TestCPU01(N)	function TestGPU(N)
tic;	tic;
X=linspace(0,2*pi,N);	gpuX=parallel.gpu.GPUArray.linspace(0,2*pi,N);
A=sin(X);	gpuA=sin(gpuX);
toc;	A=gather(gpuA);
	toc;

Table 3.4. *Example CPU and GPU codes in MATLAB*

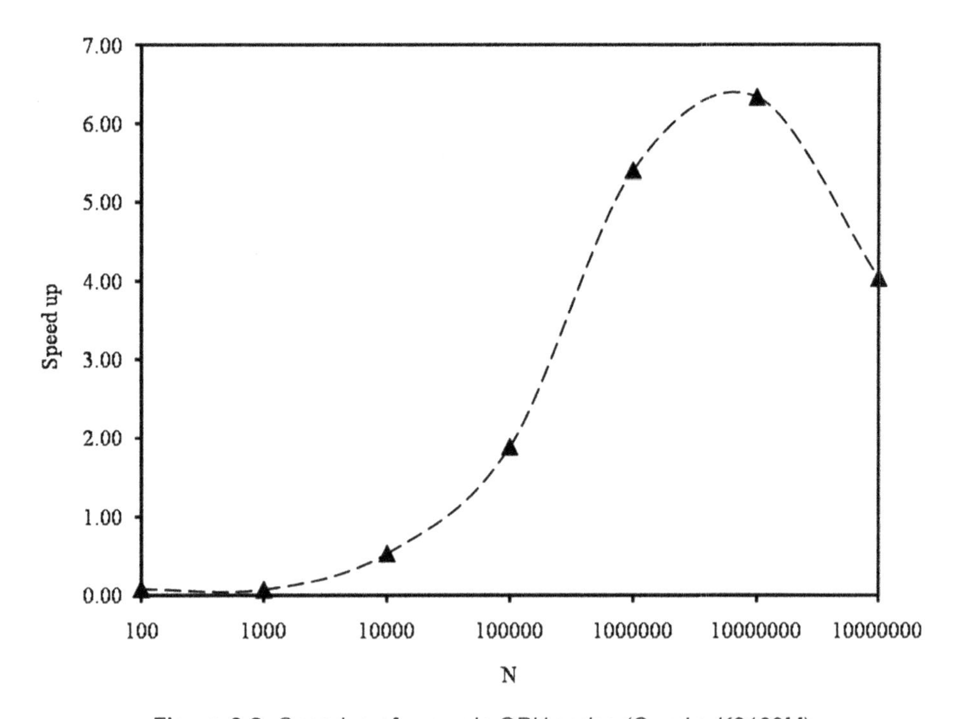

Figure 3.2. *Speedup of example GPU codes (Quadro K2100M)*

This example shows that GPU computation can improve the performance of a code in MATLAB and also verifies that the GPU functions provided in the Parallel

Computing Toolbox of MATLAB are effective; these functions will thus be used to parallelize the DICE2D code to the GPU chip. The motion update procedure of DICE2D can be directly modified into the GPU codes. For example, the position update of the particles in DICE2D and the corresponding GPU version are shown in Table 3.5.

CPU code	GPU code
%Position Update	%Position Update
X=X+Vx*DeltT;	gpuVx=gpuArray(Vx);
Y=Y+Vy*DeltT;	gpuVy=gpuArray(Vy);
T=T+Vt*DeltT;	gpuVt=gpuArray(Vt);
	gpuX=gpuX+gpuVx*DeltT
	gpuY=gpuY+gpuVy*DeltT
	gpuT=gpuT+gpuVt*DeltT
	X=gather(gpuX)
	Y=gather(gpuY)
	T=gather(gpuT)

Table 3.5. *GPU implementation of position update calculation in DICE2D*

CPU code	GPU code
for j=1:NumP	for k=1:MAX_Pn
for k=1:MAX_Pn	if gpuTagPP_C(k)
jP=PP_C(k,j)	%From CPU to GPU
if jP>0	XX1=X(gpuPP_C(k,:));
R1=R(j)	YY1=Y(gpuPP_C(k,:));
R2=R(jP)	RR1=R(gpuPP_C(k,:));
Xdef=X(jP)-X(j)	VXX1=Vx(gpuPP_C(k,:));
Ydef=Y(jP)-Y(j)	VYY1=Vy(gpuPP_C(k,:));
...	VTT1=Vt(gpuPP_C(k,:));
end	...
end	% Calculate in GPU
end	gpuXdef=gpuXX1-gpuX;
	gpuYdef=gpuYY1-gpuY;
	...
	%From GPU to CPU
	DL=gather(gpuD);
	for j=1:NumP
	jP=PP_C(k,j);
	if jP>0
	...
	end
	end
	end

Table 3.6. *GPU implementation of the contact force calculation in DICE2D*

For some parts of DICE2D, the GPU implementation cannot be directly conducted as shown in Table 3.5; more complex treatments are required. For example, Table 3.6 shows the GPU implementation of the contact force calculation in DICE2D with modifications. Instead of processing each particle separately, the GPU code performs matrix operations for all of the particles in each contact layer. The mathematical operations such as *cos* and *sin* are calculated as matrix operations in the GPU. The majority of the computation is performed in the GPU rather than in the CPU. The GPU computation results are collected into CPU matrices for further analysis. Other parts of the DICE2D code are not converted into GPU versions due to the presence of many logical operations, which are not computed more quickly in the GPU. The algorithm of the CPU code with regard to memory operation is suitable for multi-core computers but not for GPU computation, and vice versa.

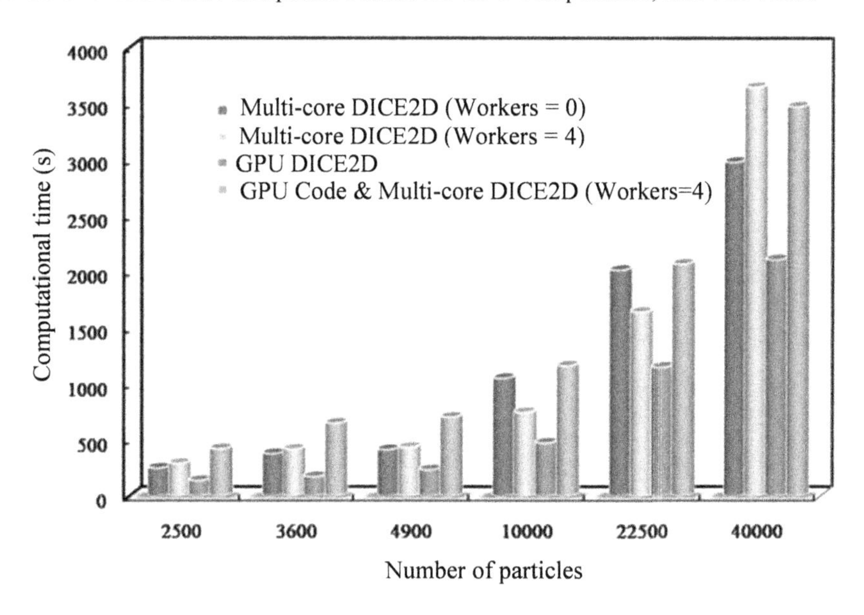

Figure 3.3. *Computational time of the multi-core DICE2D, GPU DICE2D, and GPU and multi-core DICE2D codes for the uniaxial compression test*

Owing to the presence of many logical operations in the contact detection computation, it is difficult and inefficient to implement this algorithm in the GPU; however, this algorithm can still be parallelized using *parfor*. The uniaxial compression test shown in Figure 2.3 is used to test the performance of the GPU DICE2D code. For comparison, the multi-core DICE2D using different number of workers (i.e. zero refers to the serial DICE2D), GPU DICE2D, and GPU and multi-core DICE2D (contact detection is parallelized for multi-core CPU) codes are used to solve the same problem with different number of particles. The computational time of these codes is shown in Figure 3.3. Overall, the GPU DICE2D code is completed in less time than the multi-core (i.e. parallel) DICE2D code. However, when the GPU and multi-core parallelization are mixed together, the performance is not improved, as expected, due to competition between the GPU and the multi-core functions in MATLAB (e.g. communication between the different workers influence the communication between the CPU and the GPU). This example shows that putting more computational resources together will not always improve the computational performance. Nevertheless, from this example, it can also be concluded that the GPU implementation of the DICE2D code is successful.

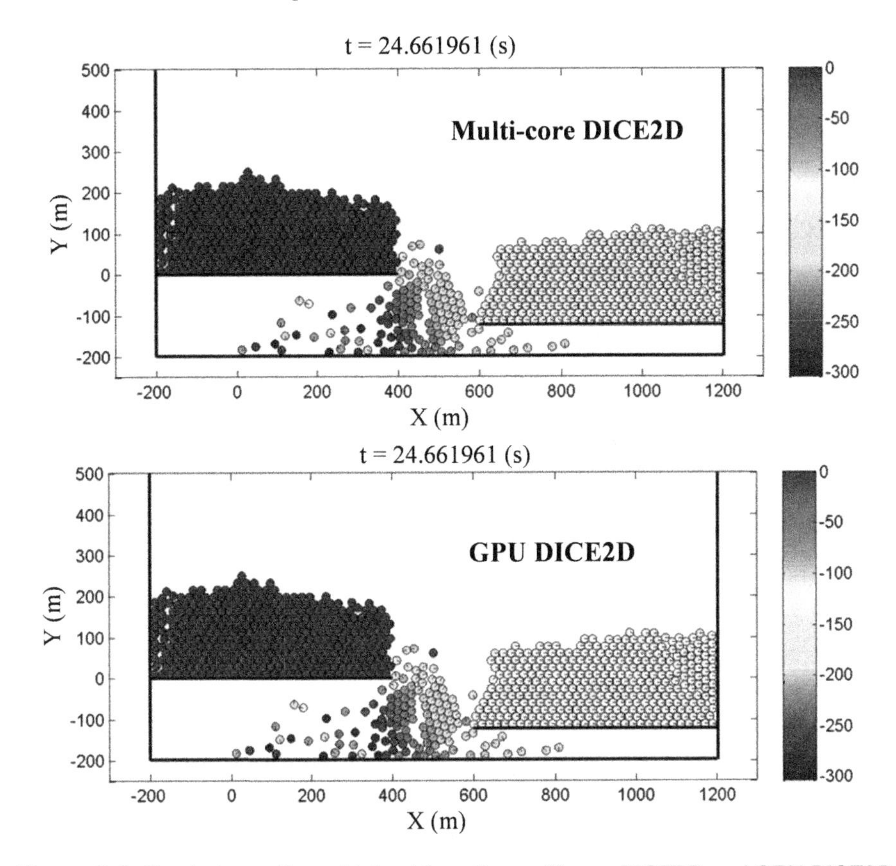

Figure 3.4. *Fracturing pattern obtained from the multi-core DICE2D and GPU DICE2D codes. For a color version of the figure, see www.iste.co.uk/zhao/computing.zip*

The simulation results of the GPU code on a fracturing pattern might be different from those of the CPU version [ZHA 12]. The long-wall mining example was solved using the GPU DICE2D code, and the simulation results are shown in Figure 3.4. It is shown that the same failure pattern is obtained,which might be because in this study, a double-precision variable is used whereas a single-precision variable was used in a study conducted by Zhao and Khalili [ZHA 12]. The GPU DICE2D code was not parallelized for the contact detection calculation; therefore, it might be inefficient for problems involving large number of contact detections, such as granular flow–like problems.

3.3. Examples

In Chapter 2, numerical examples are mainly presented to show the ability of the DICE2D code to model continuum problems. The relationship between the spring stiffness and the macro elastic parameters is presented. However, the mechanical parameters related to failure are not considered. In this section, a number of numerical examples of failure are presented to study the ability of the DICE2D code to model direct tension failure, indirect tension failure, uniaxial compression failure and triaxial compression failure.

3.3.1. *Uniaxial tension test*

Tension failure is the most common and simple failure type observed in both nature and engineering. In DEM, the tensile strength is the first strength parameter that must be calibrated. Owing to the limitations of available experimental data with regard to uniaxial tension tests, many researchers have used the indirect tension test to find the micro tension parameter corresponding to the macro experimental value (e.g. [KAZ 13]); however, in some studies, the uniaxial tension test was adopted (e.g. [SCH 13]). The author believes that the uniaxial tension test performs better than the indirect tension test because the deformation produced by the uniaxial tension test is under pure tension, whereas the indirect tension test cannot satisfy this condition. Therefore, in this example, the uniaxial tension test is adopted to study the tensile strength in the DICE2D modeling. As shown in Figure 3.5, the specimen is placed between two loading plates; the bottom plate is fixed during the calculation, whereas the upper plate is moved with a given constant velocity. The reaction force on the upper plate is recorded during the test to calculate the axial stress.

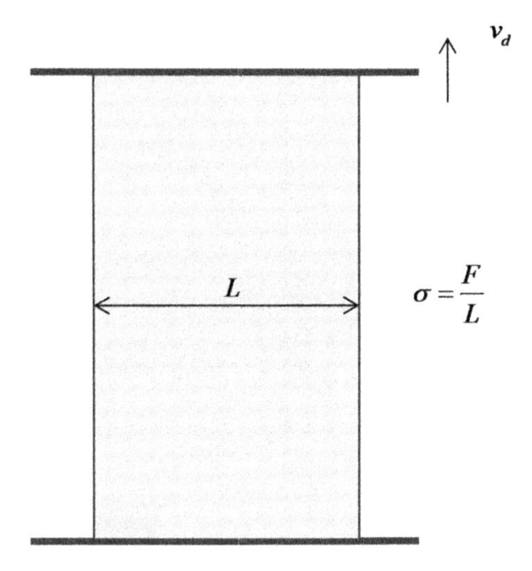

Figure 3.5. *Computational model of the uniaxial tension test*

Considering the characteristics of the tension loading conditions and the DEM, the uniaxial tensile strength of a DEM specimen can be estimated by:

$$\sigma_t = \tilde{\alpha}\frac{F_t}{\bar{D}} \quad [3.1]$$

where σ_t is the uniaxial tensional strength of the DEM, F_t is the tensile strength of a normal spring, $\bar{D}$ is the mean particle diameter of the DEM and $\tilde{\alpha}$ is a correction coefficient close to 1.

Two DEMs with different sizes are simulated to investigate the influence of the shape of the specimen on the macro tensile strength (Figure 3.6). The large specimen has a dimension of 400 m × 695.5 m. Physically, it would be difficult to build such a large model; however, in the numerical realm, it is possible to model such large specimens. Because the primary purpose of this modeling is to discover the relationship between the micro tensile parameter of the DEM and its macro tensile strength, the dimension has no influence on the results. To confirm this assumption, a much smaller model is also prepared (Figure 3.6(b)). The model parameters used in these simulations are shown in Table 3.7.

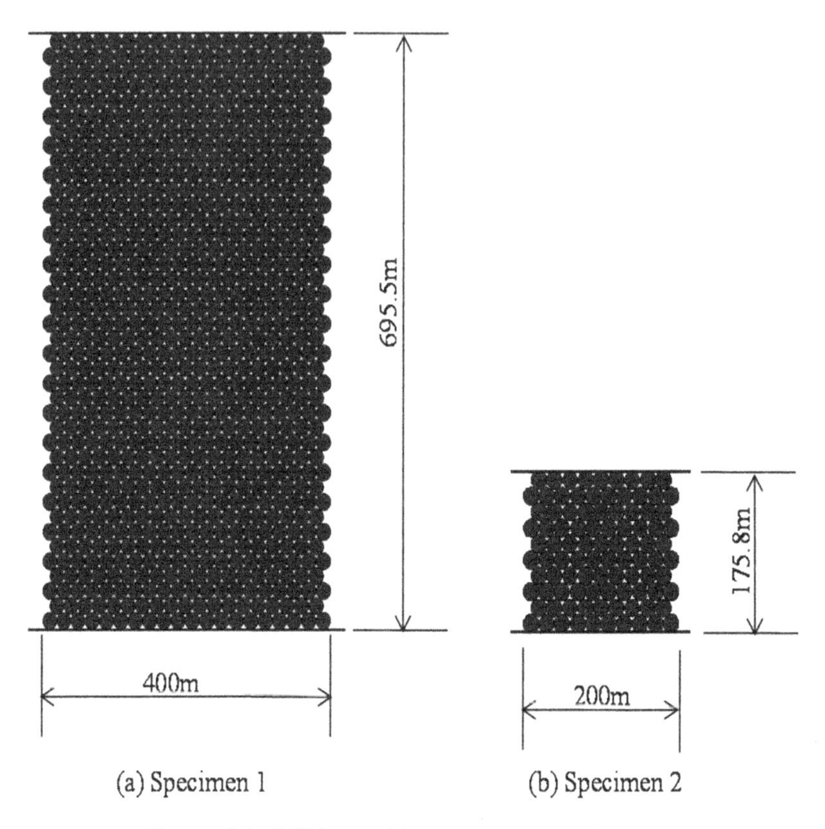

Figure 3.6. *DEMs used in the uniaxial tension tests*

Number of particles	95/779	Normal viscous coefficient (s)	0.0001
Mean particle size (m)	10	Friction angle	80
Density (kg/m^3)	1,000	Local damping	0.8
Normal stiffness (N/m)	1e8	Time step reduction factor	0.1
Shear stiffness (N/m)	–	Total steps	20,000
Tensile strength (N)	2e7	Gravity acceleration (m/s^2)	0
Cohesion	–		

Table 3.7. *Model parameters of the uniaxial tension test*

Figure 3.7 shows the failure patterns of these two specimens; similar failure patterns are observed. The stress–strain curves are shown in Figure 3.8. The modeling results indicate that the failure patterns and tensile strengths are not influenced by the dimensions or the height/width ratio of the specimens.

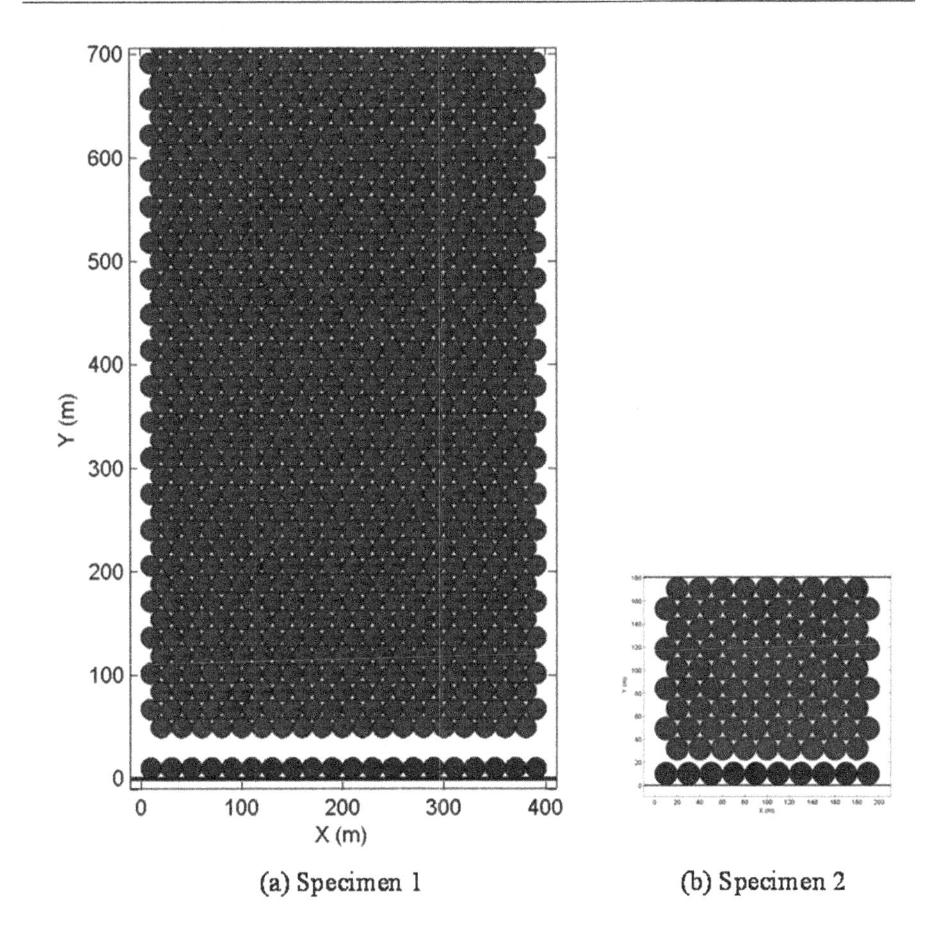

(a) Specimen 1 (b) Specimen 2

Figure 3.7. *Failure patterns of the DEMs with different sizes*

The strength of a regular triangle-packed specimen is approximately 3 MPa. The correction coefficient $\tilde{\alpha}$ in equation [3.1] is approximately 1.5. To investigate the influence of particle packing on the tensile strength, an irregular-packed specimen is used (Figure 3.9). Similar to the example given in section 2.5.1, the CN of the specimen can be adjusted using different thresholds during bond formation. Figure 3.10 shows the failure patterns of these two specimens and indicates that the CN does influence the failure pattern. For a low-CN specimen, a straight fracture is shown; however, a zigzag-type fracture is observed in the specimen with a high CN.

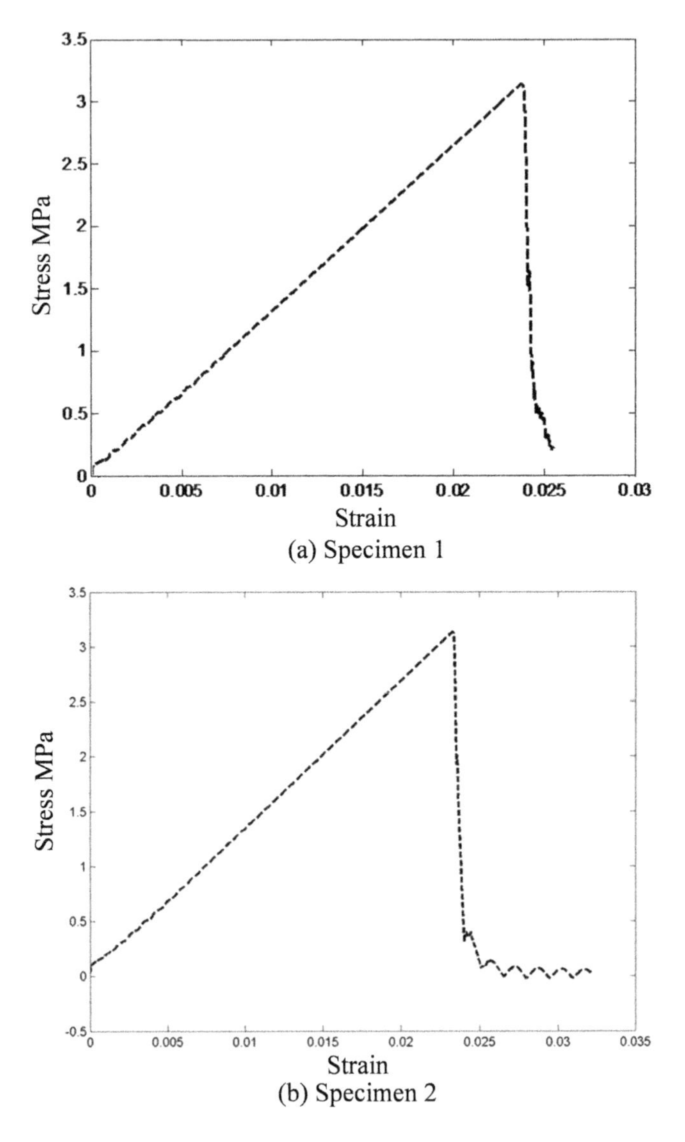

Figure 3.8. *Stress–strain curves of the DEMs of the uniaxial tension tests using different specimens*

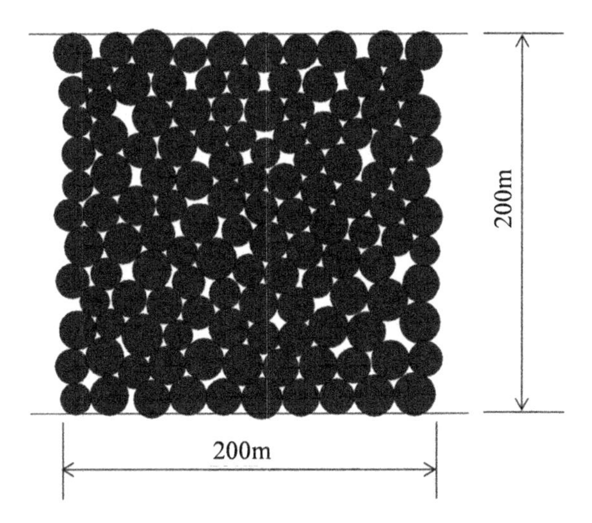

Figure 3.9. *Irregular particle packing specimen for the uniaxial tension test*

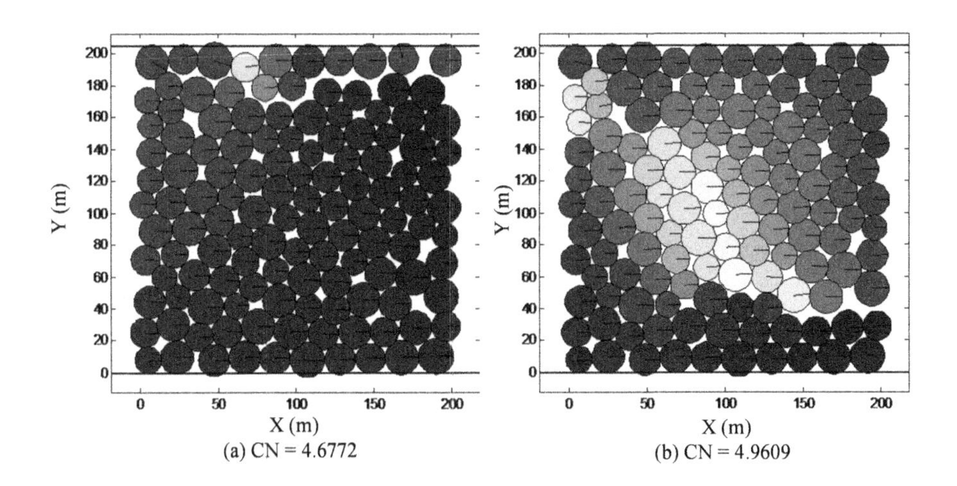

Figure 3.10. *Failure patterns of irregular-packed specimens with different CNs. For a color version of the figure, see www.iste.co.uk/zhao/computing.zip*

The stress–strain curves are shown in Figure 3.11. The low-CN specimen has a tensile strength of approximately 2 MPa, whereas the high-CN specimen shows a tensile strength of 2.5 MPa. The corresponding correction coefficient $\tilde{\alpha}$ of equation [3.1] is equal to 1.0 and 1.25, respectively. In real applications, $\tilde{\alpha}$ can be estimated to be 1.0 for loosely packed specimens, 1.25 for closely packed specimens and 1.50 for very closely packed specimens.

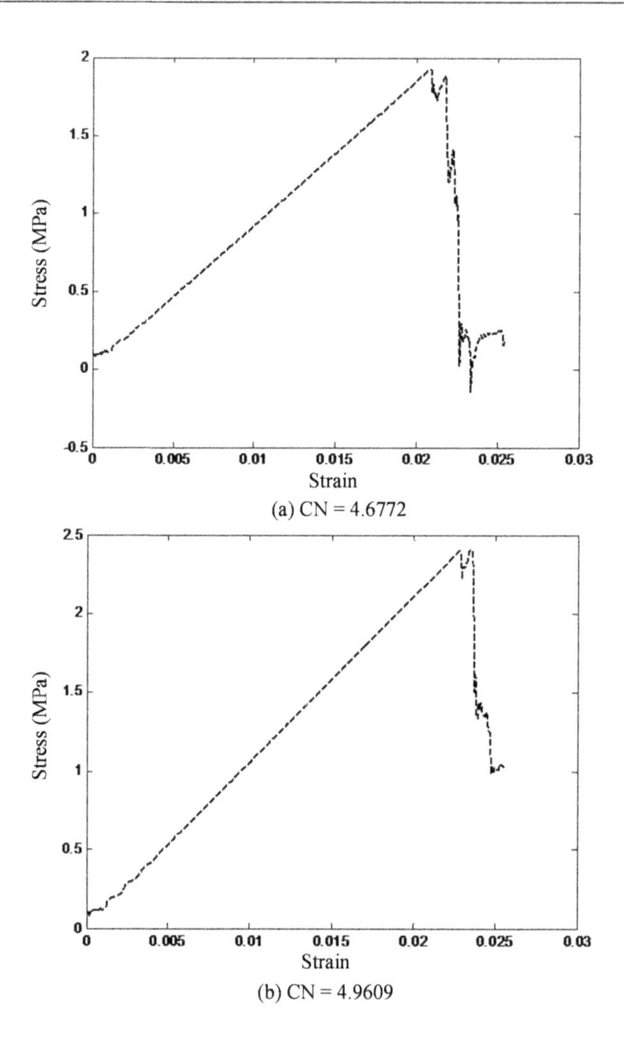

Figure 3.11. *Stress–strain curves of irregular-packed specimens with different CNs*

The influence of other parameters, such as the WRT, the K_s/K_n, the cohesion of the bond and the friction of the bond, is studied. The regular triangle-packed model shown in Figure 3.6(b) is used to reduce computational time. A total of 28 numerical tests are performed, and the model parameters used are the same as those listed in Table 3.7. The modified parameters except for the friction angle are dimensionless. Figures 3.12–3.15 show the final simulation results and reveal that both the WRT and the K_s/K_n influence the tensile strength of the model slightly. The maximum difference is shown to be approximately 10%. Therefore, equation [3.1] can be used only for estimation purposes.

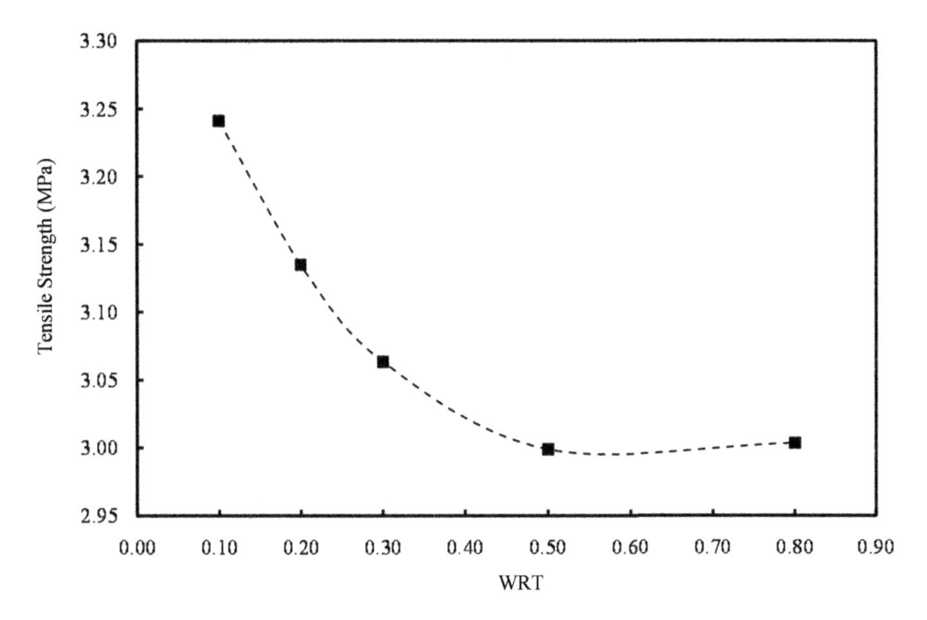

Figure 3.12. *Influence of the bond thickness ratio coefficient on the uniaxial tensile strength*

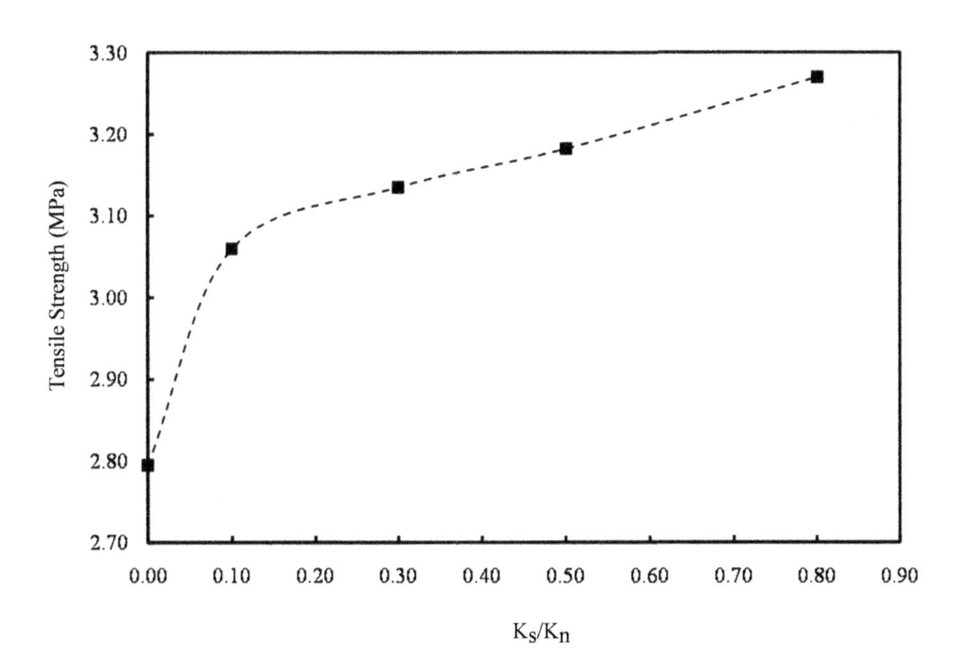

Figure 3.13. *Influence of the shear stiffness ratio on the uniaxial tensile strength*

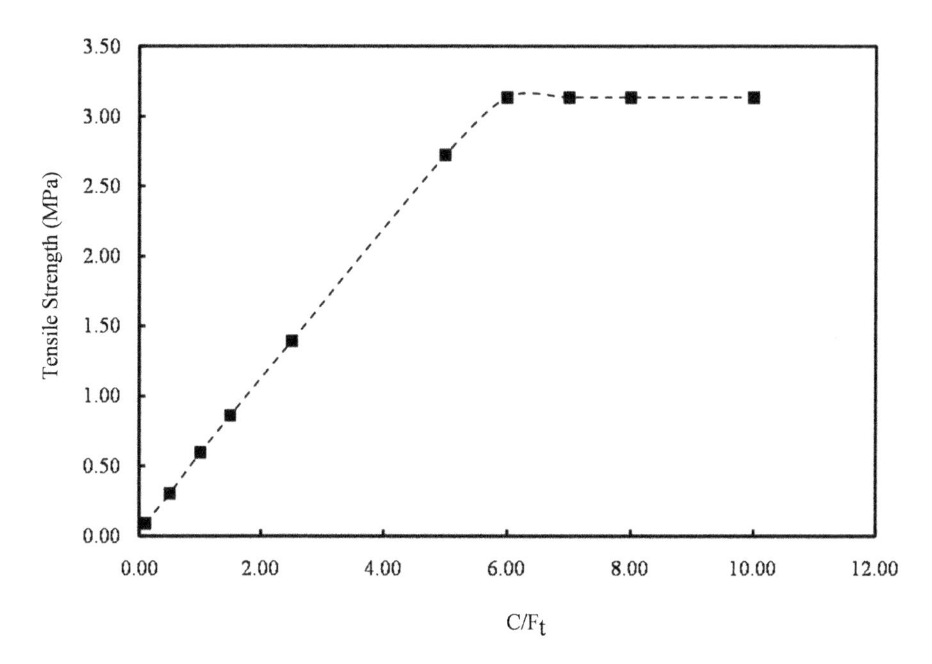

Figure 3.14. *Influence of the cohesion of the spring bond on the uniaxial tensile strength*

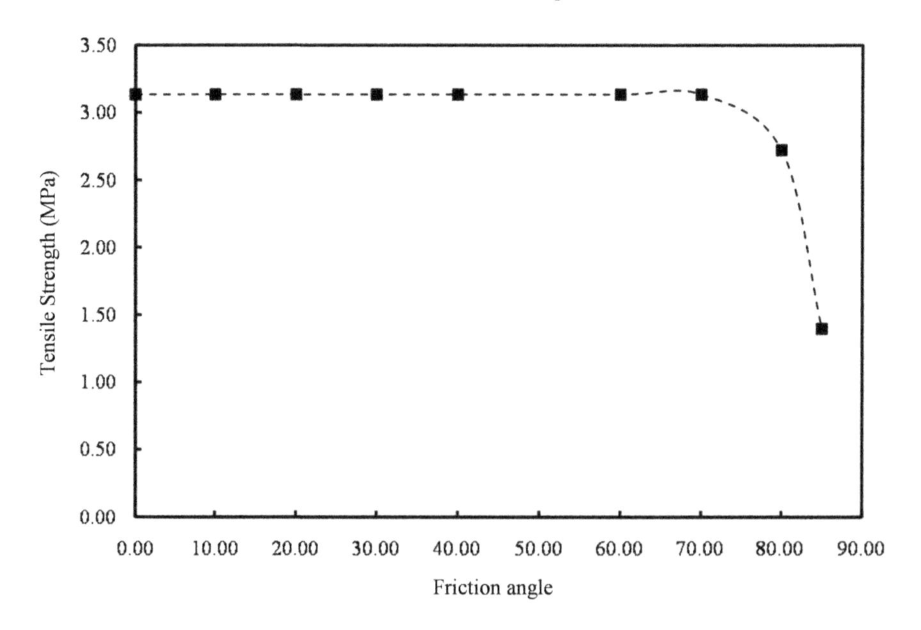

Figure 3.15. *Influence of the friction angle of the spring bond on the uniaxial tensile strength*

From Figures 3.14 and 3.15, it can be concluded that under certain conditions, the tensile strength of a DEM specimen is not influenced by the cohesion and friction angle of the contact bond. However, if the cohesion or friction angle is less than or greater than the critical value in which the tension cut condition will not be satisfied, the bond might fail under shear based on the micro Mohr–Coulomb model. For most brittle materials, tension failure is the primary failure mode. The parameters shown in Figures 3.14 and 3.15 can be used to provide additional constraints to the selection of the friction angle and the cohesion parameters for a spring bond.

3.3.2. *Brazilian disk test*

Owing to the convenience of its specimen preparation, the Brazilian disk test has become a widely used test method in rock mechanics to determine the tensile strength of a material, although there are still many debates over the validity of the testing results [LI 13]. As shown in Figure 3.16, the numerical experimental setup of the Brazilian disk test is simple: there are two loading plates and a disk specimen (i.e. cylindrical specimen in a real test). During the test, the reaction force of the upper plate is recorded, and the tensile strength is obtained using the following equation:

$$\sigma_t = \frac{2F_{\max}}{\pi DL} \qquad [3.2]$$

where L is the length/thickness of the specimen, D is the diameter and $F_{\max}$ is the peak value of the loading force. The adopted DEM is shown in Figure 3.17. The diameter is 400 m, which is large compared with a real experiment. The primary purpose is to obtain the relative relationship between the indirect tensile strength of the DEM and the uniaxial tensile strength. Therefore, the influence of the dimensions of the specimen can be ignored. Two walls are used to simulate the loading plates. The material parameters of the specimen are considered to be the same as those in the previous example. Two threshold values are used to produce specimens with different CNs. Figure 3.18 shows the failure patterns of these two specimens; the failure pattern of the specimen is shown to be significantly influenced by the CN. For the specimen with a low CN, the failure zone is at the bottom, whereas for the specimen with a high CN it is at the top. The fracture surface also appears straighter when the CN is high. The loading curves are also shown in Figure 3.19, which shows that a higher strength is obtained for specimens with higher CNs. The corresponding indirect tensile strengths of the specimens are found to be 1.00 and 1.23 MPa, respectively, which are approximately half of the uniaxial tensile strength of the materials considered. On the basis of the work of Li and Wong [LI 13], these values represent the lowest of the experimentally observed values of the ratio between the indirect and the direct tensile strengths (e.g. 0.5–2.0).

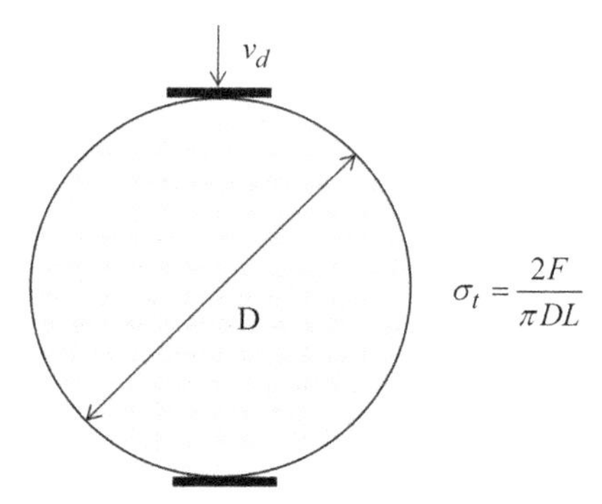

Figure 3.16. *Geometry model and loading condition of a Brazilian disk test*

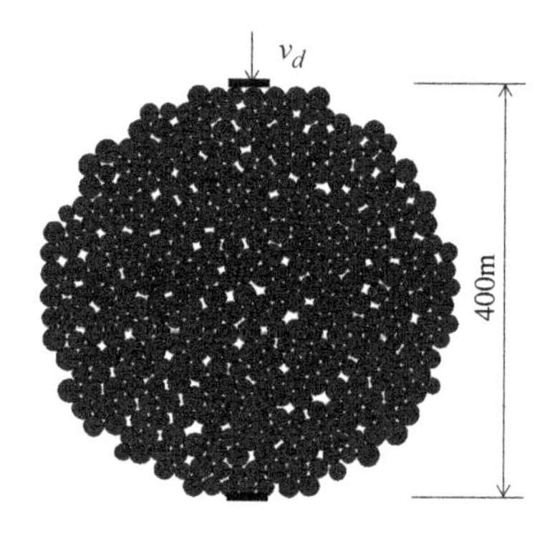

Figure 3.17. *DEM of the Brazilian disk test*

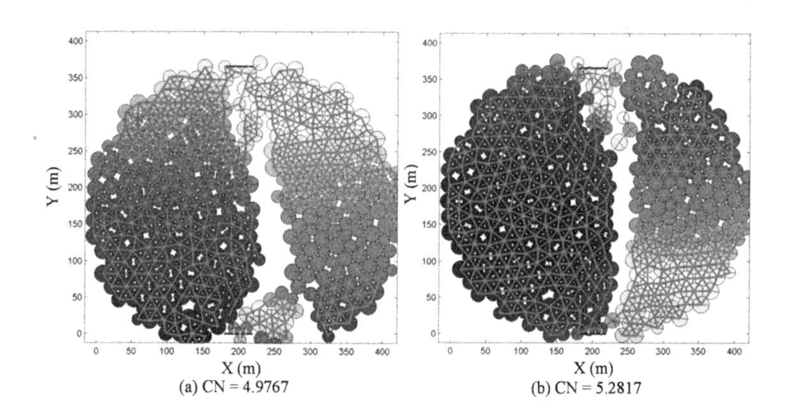

Figure 3.18. *Failure pattern of the specimen with different CNs. For a color version of the figure, see www.iste.co.uk/zhao/computing.zip*

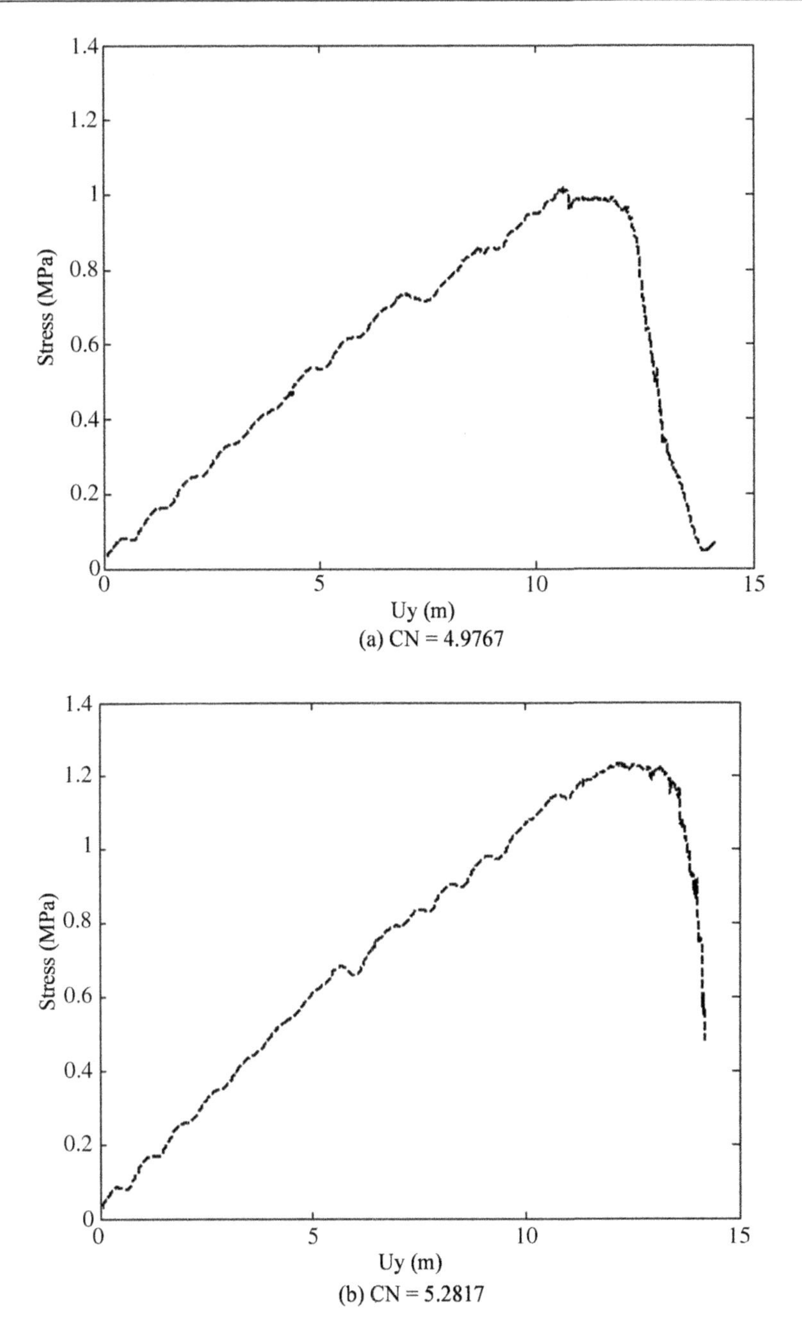

Figure 3.19. *Stress–strain curves of the Brazilian disk test*

The effects of the micro parameters on the indirect tensile strength are determined to describe whether the indirect tensile strength will appear with the same tendency as it occurs when the micro parameters change. Figures 3.20 and 3.21 show the influence of the WRT and K_s/K_n on the indirect tensile strength. With an increase of the WRT, the indirect tensile strength decreases. Compared to Figures 3.12 and 3.13, it can be concluded that the indirect tensile strength can reflect a change in the WRT and K_s/K_n in the same way as does the uniaxial tensile strength. Therefore, the indirect tensile strength can still be used to calibrate the tensile strength; however, in DEM simulations, this difference should be considered.

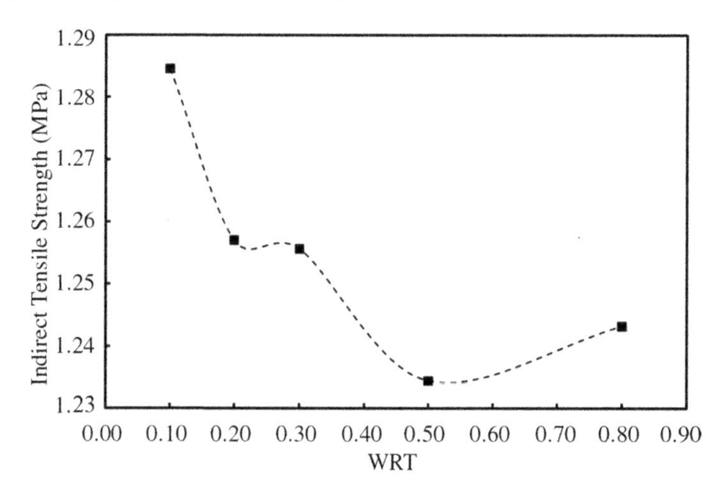

Figure 3.20. *Influence of the bond thickness ratio on the indirect tensile strength*

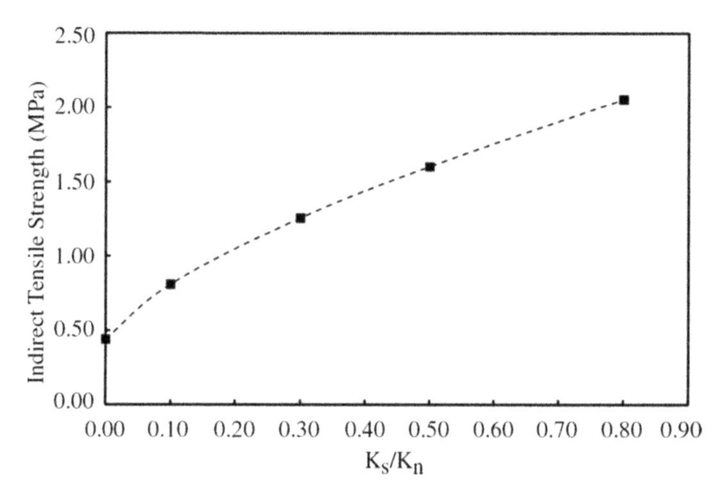

Figure 3.21. *Influence of the shear stiffness ratio on the indirect tensile strength*

3.3.3. *Uniaxial compressive strength test*

The uniaxial compressive strength (UCS) test is another widely used calibration test in DEM simulations. In this example, the GPU DICE2D code is used to model a UCS test. The experimental setup and the DEM are shown in Figure 3.22. The irregular-packed specimen with a CN of 5.0788 is used. Model parameters are selected to be the same as those in the previous example. The simulation results are shown in Figure 3.23.

The failure pattern of the specimen is commonly observed in the UCS test (see Figure 3.23). The simulated compressive strength is 8 MPa. The ratio between the compressive strength and the indirect tensile strength is found to be approximately 7, whereas the ratio between the compressive strength and the uniaxial tensile strength is found to be in the range of 3–4. Therefore, the ratio problem in the DEM code still exists in the DICE2D code which can be solved using the clump or cluster particle scheme. Considering the Poisson's ratio and the current results, the solution provided by Scholtes and Donze [SCH 13] seems to not fully solve the ratio problem of the DEM code; for example increasing the compressive-to-tensile-strength ratio to a realistic value using a high-CN specimen might result in the Poisson ratio limitation problem (Figure 2.13).

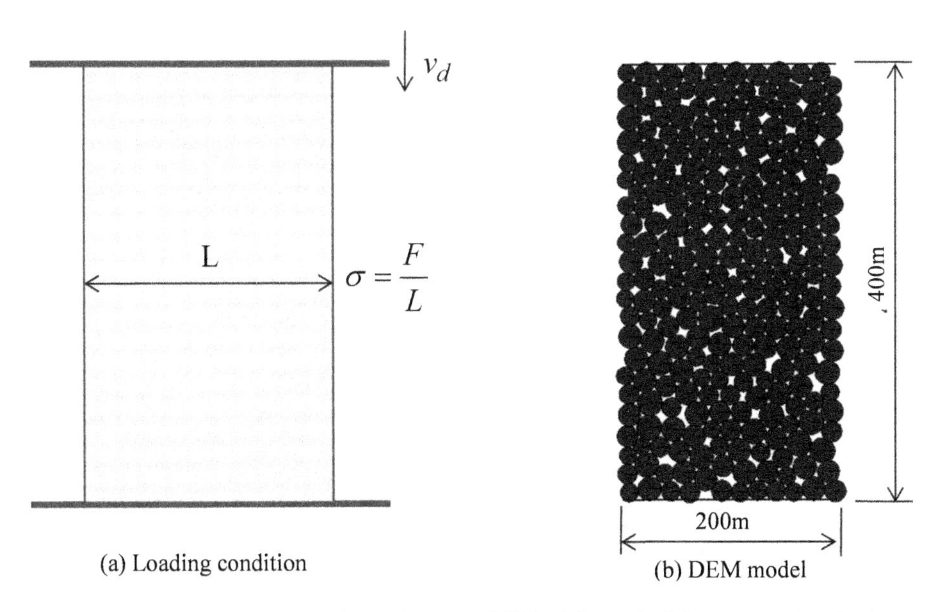

(a) Loading condition

(b) DEM model

Figure 3.22. *Loading configuration and DEM of the uniaxial compression test*

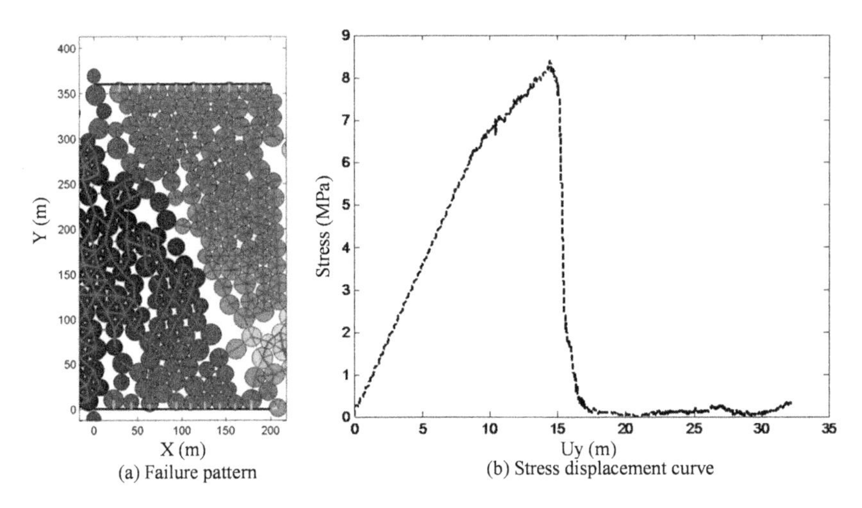

Figure 3.23. *Failure pattern of the specimen and loading curve of the UCS test. For a color version of the figure, see www.iste.co.uk/zhao/computing.zip*

The parameter analysis of the UCS test is shown in Figures 3.24 and 3.25. As expected, the compressive strength increases with the increase in friction angle; bond cohesion is also shown to first increase with the cohesion and reach a stable value (Figure 3.25). Therefore, if the friction angle is determined, the compressive strength can be used to determine the cohesion.

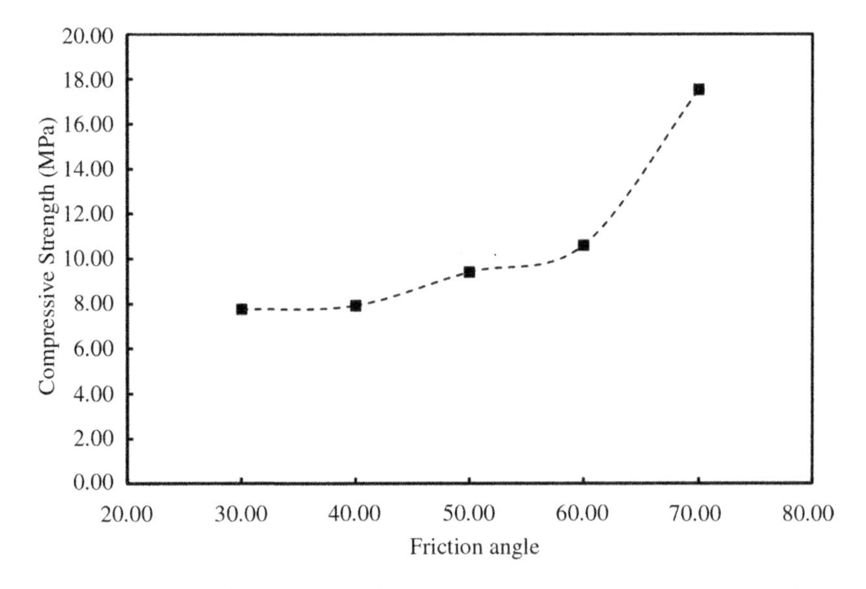

Figure 3.24. *Influence of the friction angle on the compressive strength*

3.3.4. *Triaxial compressive test*

The triaxial compression test (Figure 3.26) is simulated to obtain the relationship between the input micro friction angle and the macro observed friction angle. To apply a confining pressure onto the walls of the specimen, a force boundary condition is used; the idea behind this approach is to give the wall with mass and allow it to affect the calculations in Newton's second law of motion. During these calculations, the mass walls are treated like flat particles. The same particle model as that used in the previous example was used in the triaxial tests. One modeling result is shown in Figure 3.27. With increasing confining pressure, the strength increases, and the post-peak strength becomes smooth with a residual strength, as commonly observed in real experiments. Figure 3.28 shows the failure patterns of the specimens with different confining pressures, which are shown to influence the failure pattern of the specimen. With the results from a number of numerical tests, the failure envelope of the DEMs with different input friction angles can be obtained (Figure 3.29). The relationship between the macro friction and micro friction angles is shown in Figure 3.30. It can be found that in DEM simulations, the input friction angle does not equal the macro observed friction angle; for example when a zero friction angle is used as the micro input, the macro friction angle is found to be approximately 20°.

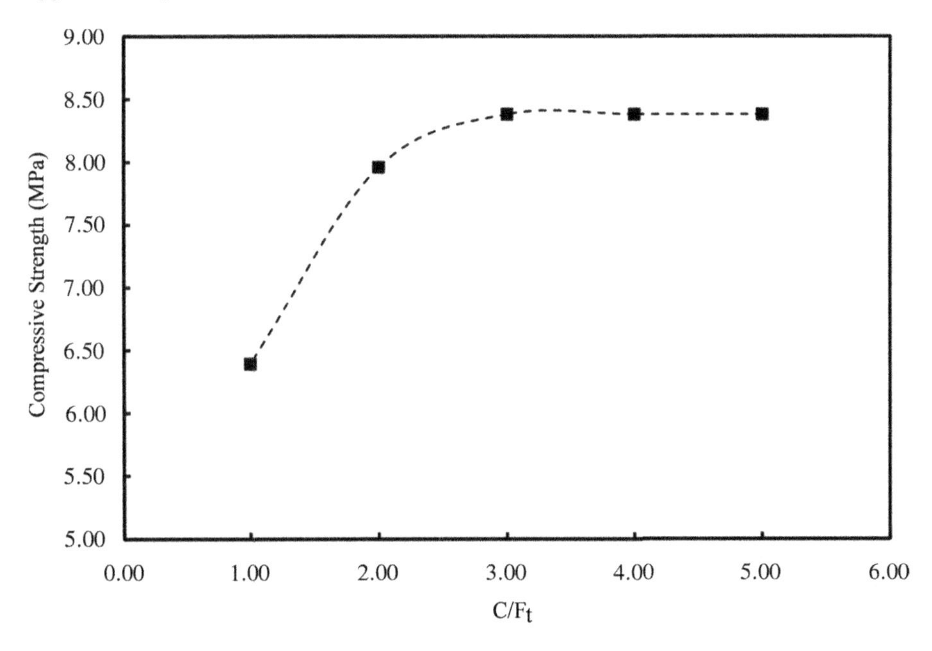

Figure 3.25. *Influence of the cohesion of spring bond on the uniaxial compressive strength*

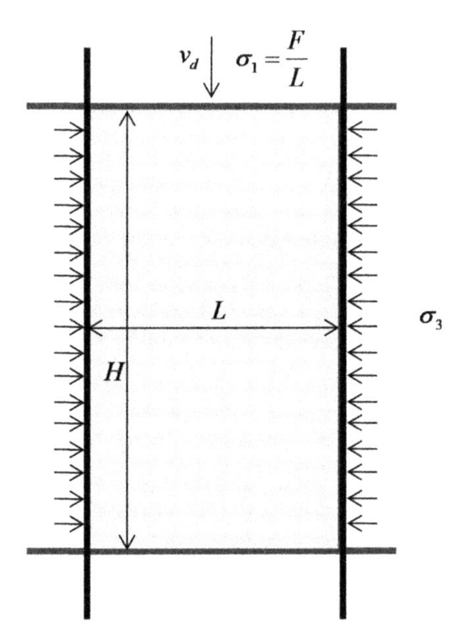

Figure 3.26. *Model configuration of the triaxial test*

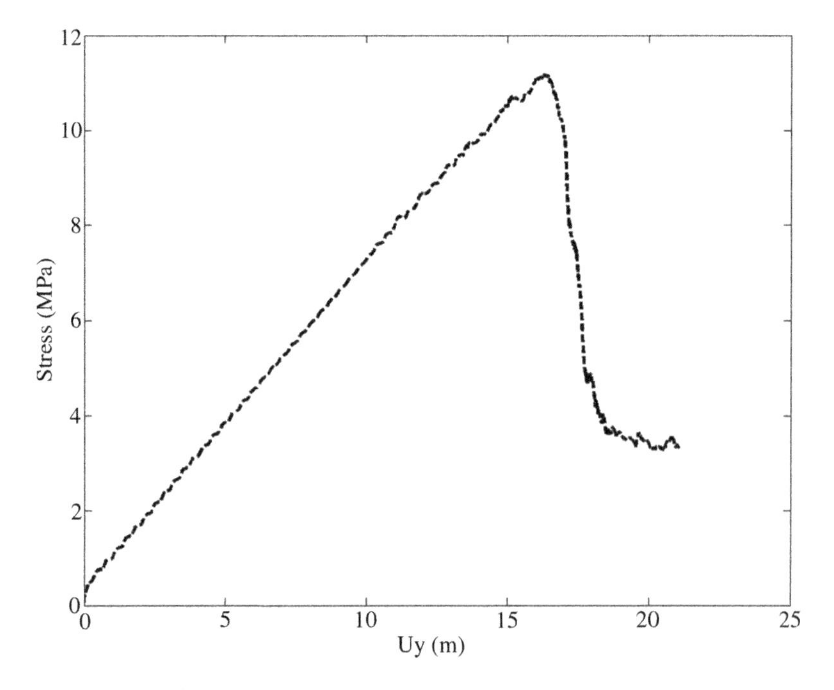

Figure 3.27. *Loading curve of the triaxial test with a confining pressure of 1.5 MPa*

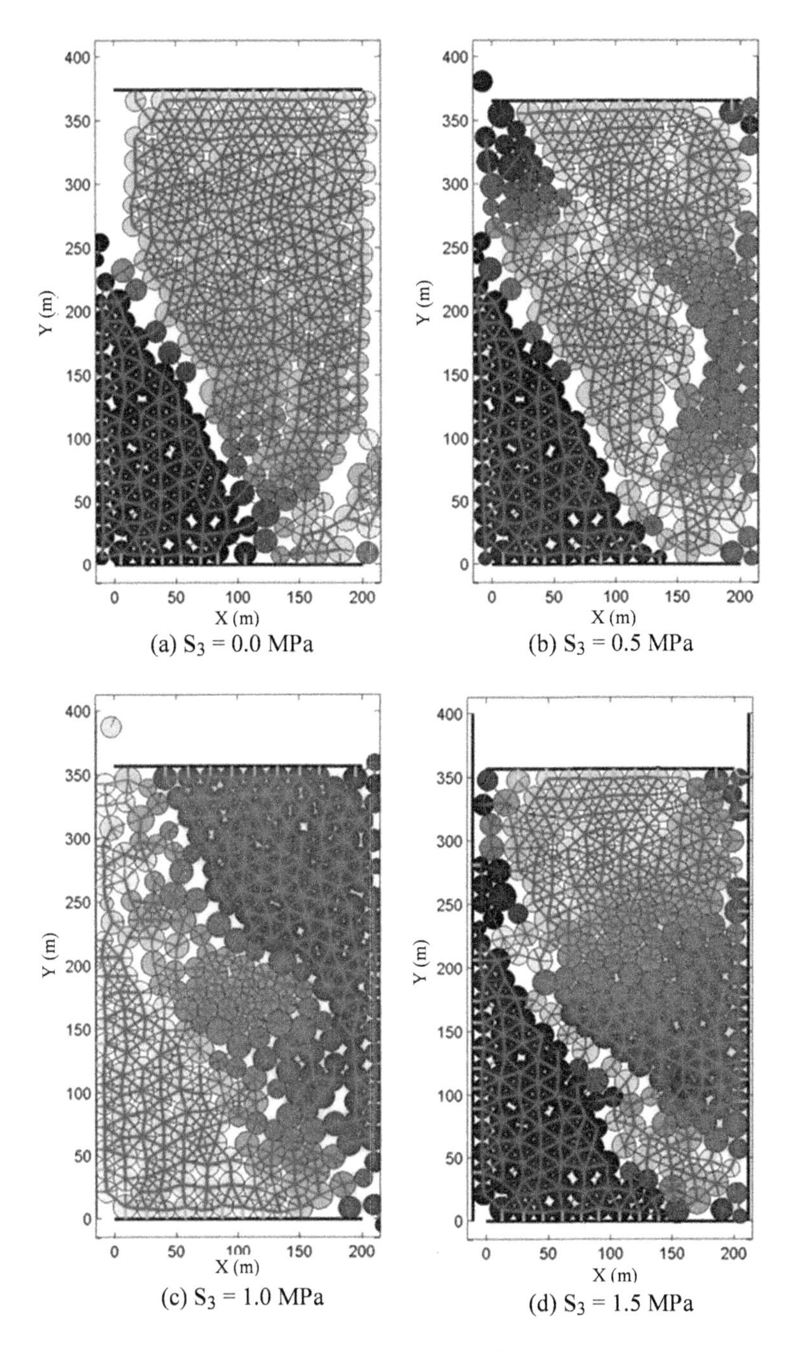

Figure 3.28. *Failure patterns of specimens under different confining pressures. For a color version of the figure, see www.iste.co.uk/zhao/computing.zip*

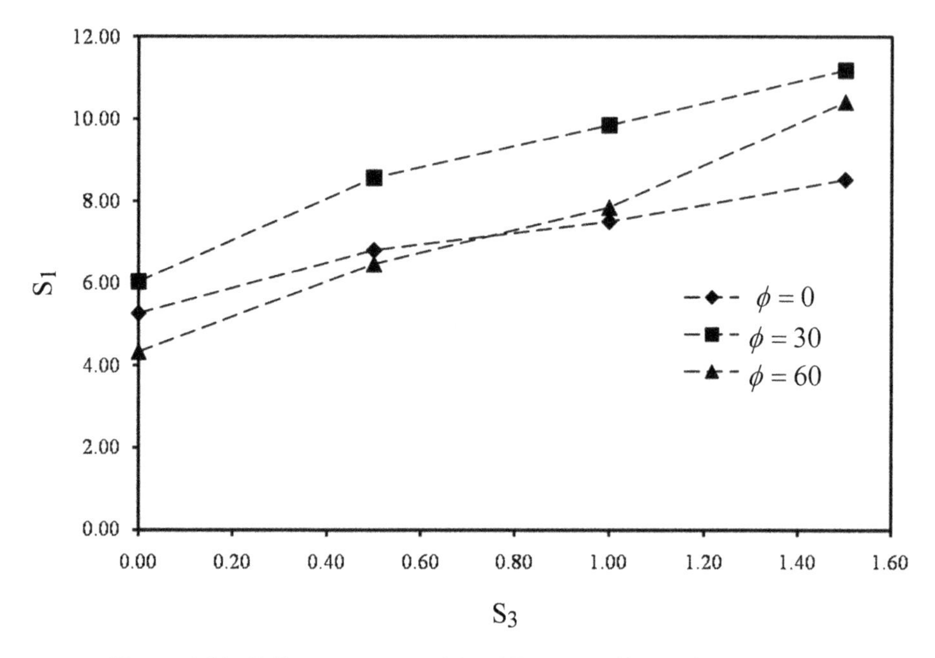

Figure 3.29. *Failure envelope of the DEM with different friction angles*

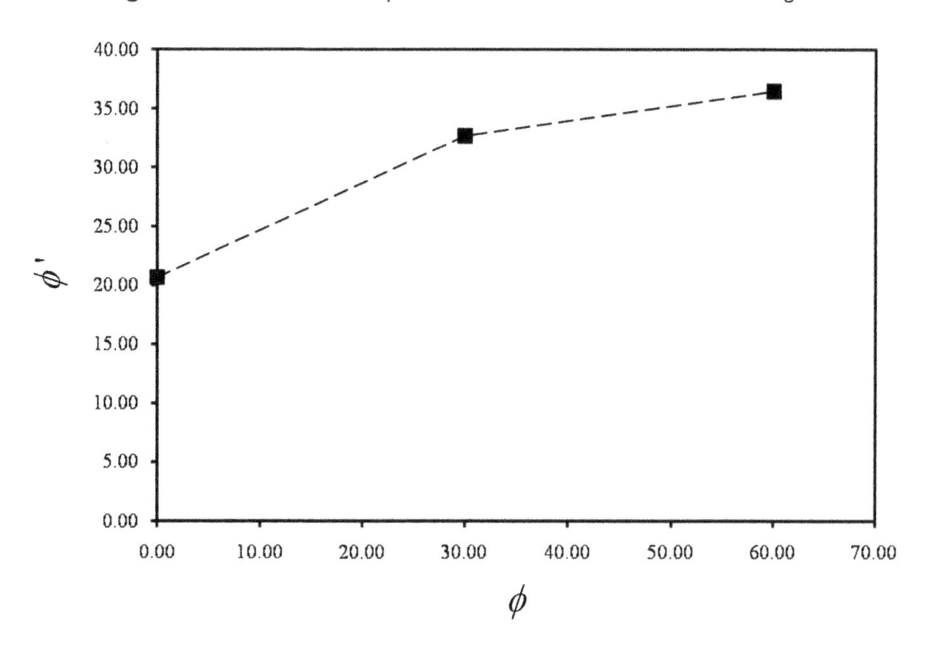

Figure 3.30. *Relationship between the input micro friction angle and the output macro friction angle of the DEM*

3.4. Conclusion

The DICE2D code is parallelized on a GPU chip by replacing certain CPU matrix operations with GPU matrix operations using the Parallel Computing Toolbox of MATLAB. From the modeling results, the GPU DICE2D code is shown to provide better performance than the multi-core DICE2D code; however, this performance improvement is not as significant as other classical GPU parallelization procedures. The reason for this outcome is that the GPU implementation using MATLAB is a high-level parallelization. Multiple numerical examples are investigated using the GPU DICE2D code to determine the relationship between the micro and macro spring failure parameters. From these examples, a calibration procedure is suggested: first, the tensile strength can be obtained using a uniaxial or indirect tension test; then, the cohesion can be obtained from a UCS test. Finally, the friction angle can be determined using TCS test data.

4

DICE2D and Cluster

A cluster is a collection of computers connected through a high-speed network that allows them to work together to solve a problem. The newly developed technologies of HPC (e.g. multi-core CPUs and GPUs) can be easily integrated into a cluster using a modular concept. A cluster usually uses a distributed memory system; therefore, model decomposition and communications between different processors (i.e. computer nodes) are required to be handled explicitly in the parallel code for a cluster. In the implementation, the MPI[1] and the parallel virtual machine[2] are two commonly used programming tools. The Parallel Computing Toolbox® of MATLAB®, which is a high-level programming environment, was used for the implementation of the multi-core (parallel) DICE2D code and produced a parallel DICE2D code that was independent of the operating system and parallel hardware. In this study, the parallel DICE2D code uses the computational resources of a moderate-sized cluster to simulate certain rock engineering problems. In Chapter 3, the GPU DICE2D code showed better performance than the parallel DICE2D code (Figure 3.3). In this Chapter, the parallel DICE2D code runs on a moderate-sized cluster called "Leonardi". The results show that the performance of the parallel DICE2D code is significantly influenced by the hardware platform. There are some negative comments on the Parallel Computing Toolbox of MATLAB; for example a comment[3] from MIT in 2011 indicated that it was "more trouble than it is worth". From the evidence shown in this chapter, a more positive review might be concluded.

---

1 http://www.mpi-forum.org/ (accessed on 01-03-2015).

2 http://www.csm.ornl.gov/pvm/ (accessed on 01-03-2015).

3 http://tig.csail.mit.edu/wiki/TIG/DistributedComputingWithMATLAB (accessed on 01-03-2015).

4.1. Leonardi cluster

The Leonardi cluster at UNSW consists of 2,944 AMD Opteron 6174 2.20-GHz processors. The cluster has a total of 5.8-TB physical memory and a 100-TB hard disk storage (each core has approximately 2 GB memory). MATLAB2014b is available on the cluster; this program is adopted as the running environment for the parallel DICE2D. The laptop accessing the cluster runs on a Windows operating system, whereas the cluster uses a Linux operating system. To connect to the cluster from the laptop, PuTTY software is used. Figure 4.1 shows the interface of PuTTY. Instructions for using PuTTY are available at www.putty.org. A quick operation of PuTTY is explained as follows. The first step is to input the name of the cluster, for example leonardi.eng.unsw.edu.au, into the text box for the host name; the port number is set to 22 as default (see Figure 4.1). Then, the *Open* button is clicked, and a command window will popup. The user name and password are entered. Finally, PuTTY will access the head node of the cluster. A front message (username@hostname) will be displayed to indicate the success of the connection. However, computations cannot be performed on the head node, and computing resources must be allocated using a Portable Batch System (PBS) job script.

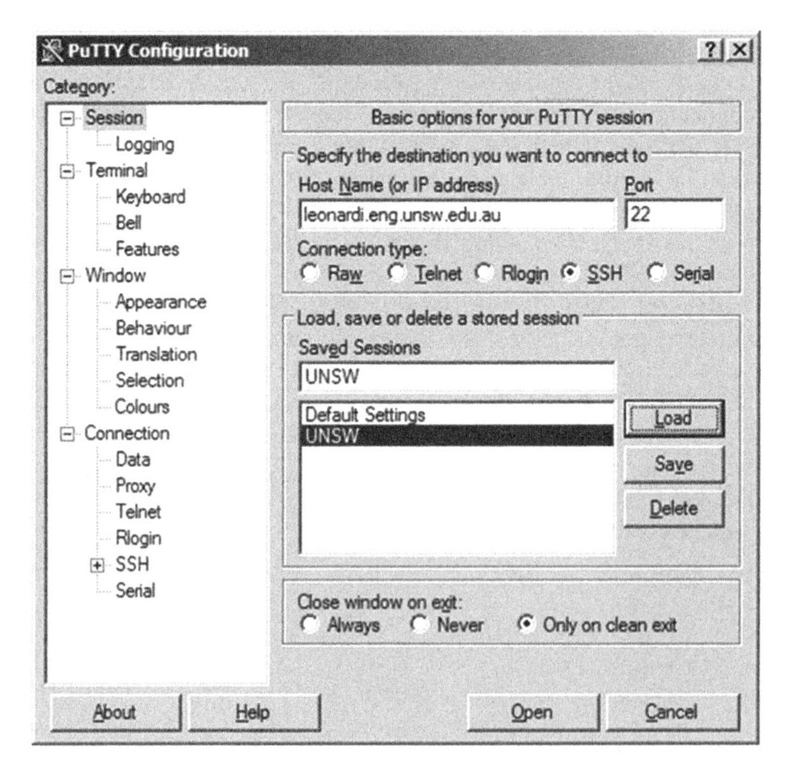

Figure 4.1. *PuTTY interface*

4.2. Run DICE2D on cluster

To run the program on the cluster, the first step is to upload the parallel DICE2D folder to the cluster. WinSCP[4] is used for this step. After the parallel DICE2D is uploaded to the cluster, the following command is run (Figure 4.2):

qsub -l nodes=1:ppn=48,mem=90gb,vmem=90gb,walltime=0:30:00 -q debug -I

This command requests a computer node with 48 cores together with physical memory of 90 GB and virtual memory of 90 GB. The requested computational time (wall time) is one half-hour. The type of queue requested is a debug queue.

When the computational resources are allocated successfully, the cluster will display a message (e.g. *job 46041.leonardi.eng.unsw.edu.au ready*) (see Figure 4.2). The name of the allocated computer node will replace the head node name. For example the allocated node in Figure 4.2 is ec01b01. Before running the parallel DICE2D, the module should be confirmed to be present on the cluster using the command: *module available*. If MATLAB is in the list, then the next step is to load MATLAB into the work environment. To load MATLAB, we use the *module load* command. Finally, copyright information for MATLAB will be displayed as in Figure 4.2. The MATLAB scripts used to run the parallel DICE2D are also shown in Figure 4.2. The example used 10 cores for the simulation. The uniaxial compression test of a regular-packed specimen is adopted to determine the performance of the parallel DICE2D on the cluster. Two models are run: one model has 10,000 particles and the other has 40,000 particles. The computational time required for the parallel DICE2D and the GPU DICE2D are plotted in Figure 4.3. The solid line represents the computational time of the GPU DICE2D for a DEM with 10,000 particles, whereas the dashed line refers to the computational time of the DEM with 40,000 particles. Different numbers of cores are used by the parallel DICE2D on the cluster. However, because of the limitation of available memory (90 GB), the maximum number of cores is selected as 20.

From Figure 4.3, it is clear that the parallel DICE2D shows a better performance than the GPU DICE2D. The optimal number of cores is around 10. The parallel DICE2D that uses the Parallel Computing Toolbox of MATLAB performs faster than the GPU DICE2D. However, when the number of cores exceeds 10, the performance decreases with the increase in the number of cores. Therefore, parallel DICE2D still does not fully use the computational resources of a cluster. This problem might be solved using a future Parallel Computing Toolbox of MATLAB.

4 www.winscp.net

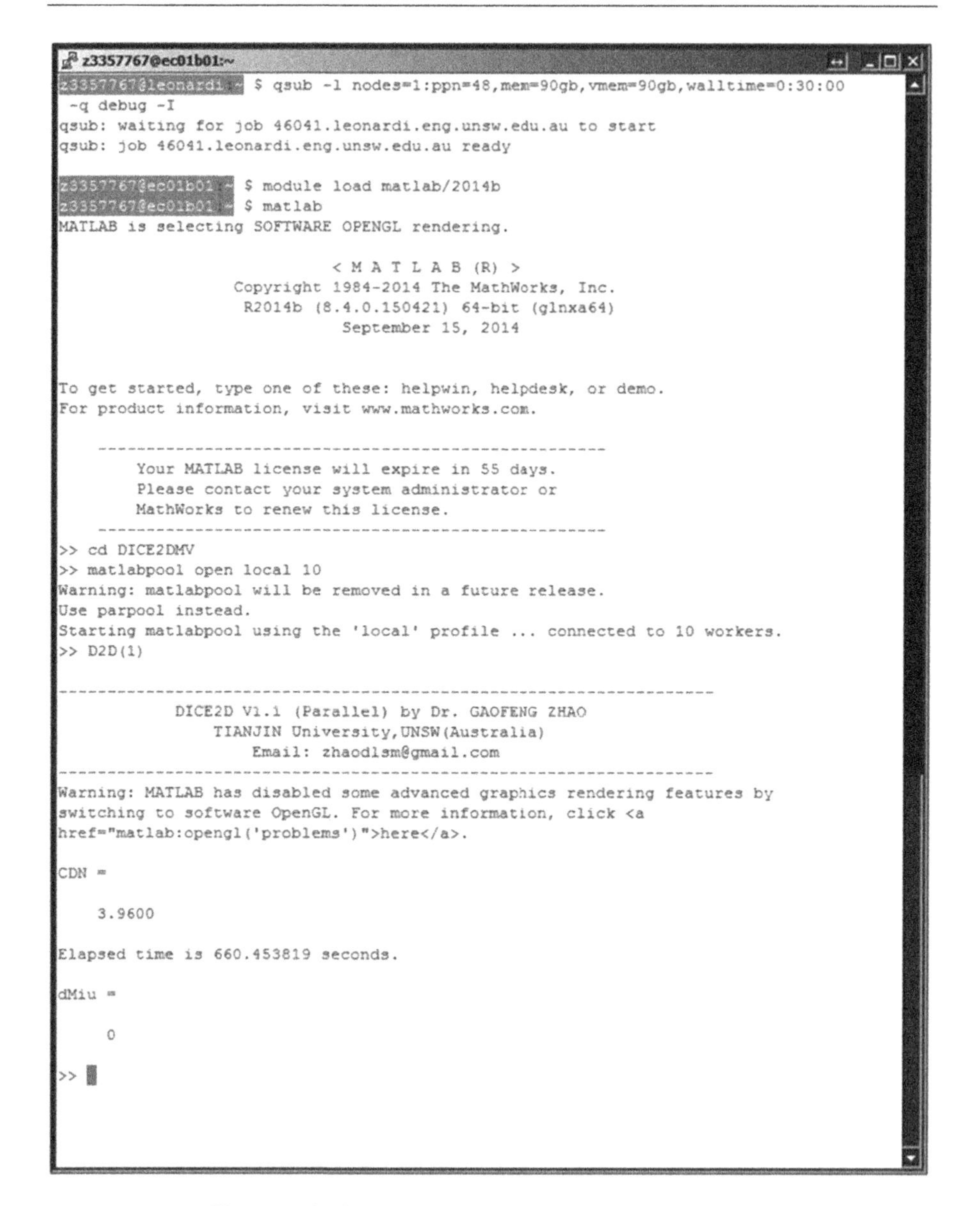

Figure 4.2. *Running the parallel DICE2D on Leonardi*

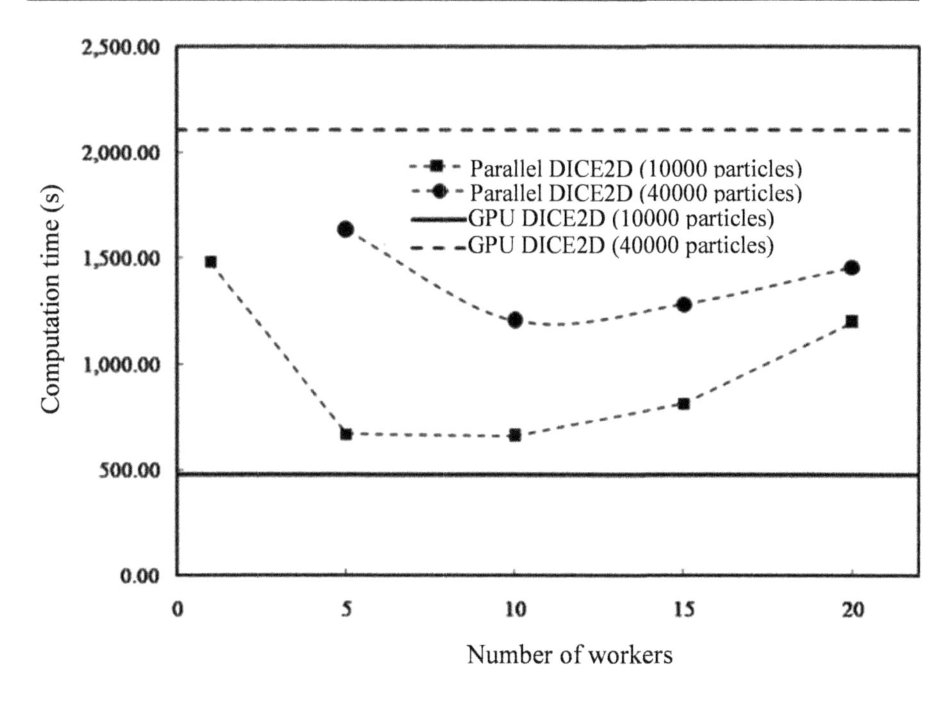

Figure 4.3. *Performance of parallel DICE2D running on a cluster*

4.3. Numerical examples

4.3.1. *Collapse of trees under gravity*

In this section, a similar problem to the tree collapse problem described in Chapter 2 is simulated. An image-based modeling technique is adopted to construct a computational model of three trees (Figure 4.4). These trees will collapse under gravity. The main purpose of the example in section 2.5.3 was to determine the ability of DICE2D on a granular flow that is dominated by a dynamic contact. In this example, the bond constitutive model is used. The WRT is set to 0.8. The corresponding model is a cohesive material model. The model and the material parameters are shown in Table 4.1. The simulation results are shown in Figure 4.5. These trees will initially undergo large deformation. Fragmentations are then formed when the deformation exceeds a certain value. In addition, these irregular fragments will move under gravity. This example shows the ability of DICE2D to model large deformations, fragmentations and irregular granular flows. These processes are typical phenomena in rock engineering applications such as block caving.

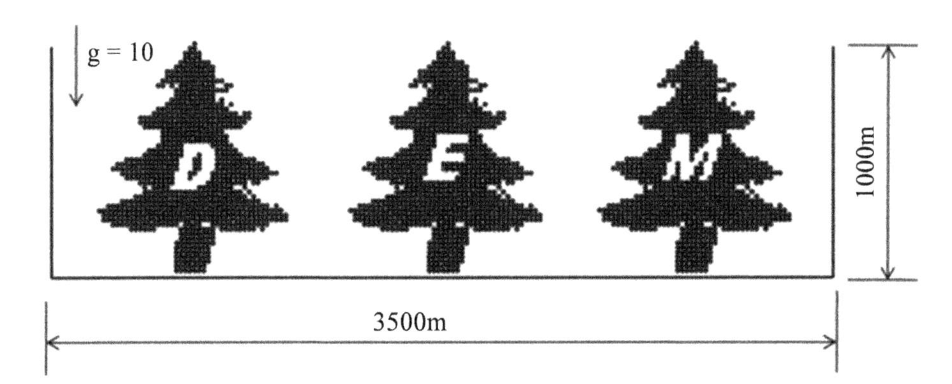

Figure 4.4. *Computational model for the three trees collapsing problem*

Number of particles	2,539	Normal viscous coefficient (s)	0.001
Mean particle size (m)	10	Friction angle	60
Density (kg/m^3)	1,000	WRT	0.8
Normal stiffness (N/m)	1e8	Local damping	0.01
Shear stiffness (N/m)	1e7	Time step reduction factor	0.1
Tension strength (N)	5e8	Total steps	10,000
Cohesion (N)	10e8	Gravitational acceleration (m/s^2)	10

Table 4.1. *Model parameters for the trees collapsing problem*

4.3.2. *Rock cutting*

Rock cutting is a typical problem that can be regarded as a sign of human civilization. In the past century, multiple studies investigated rock cutting. Theoretical models based on mechanical analyses were developed to predict the cutting force. In addition, experimental investigations were carried out to determine the essential cutting parameters. However, because of the limitations of continuum mechanics on fracturing and fragmentation analyses, these models on rock cutting remain mainly empirical. With the development of computational technology, numerical methods provide a powerful tool to study rock cutting. In this investigation, a typical example is simulated using the parallel DICE2D. The computational model is shown in Figure 4.6. A rock block of dimension 200 m × 400 m is cut by a rigid cutter. The cutter is represented by an inclined wall with an angle of 30° between the horizontal line. The cutting velocity is selected as 5 m/s and applied in the horizontal direction. The upper portion of the block is expected to be cut into rock fragments. The model and material parameters are listed in Table 4.2.

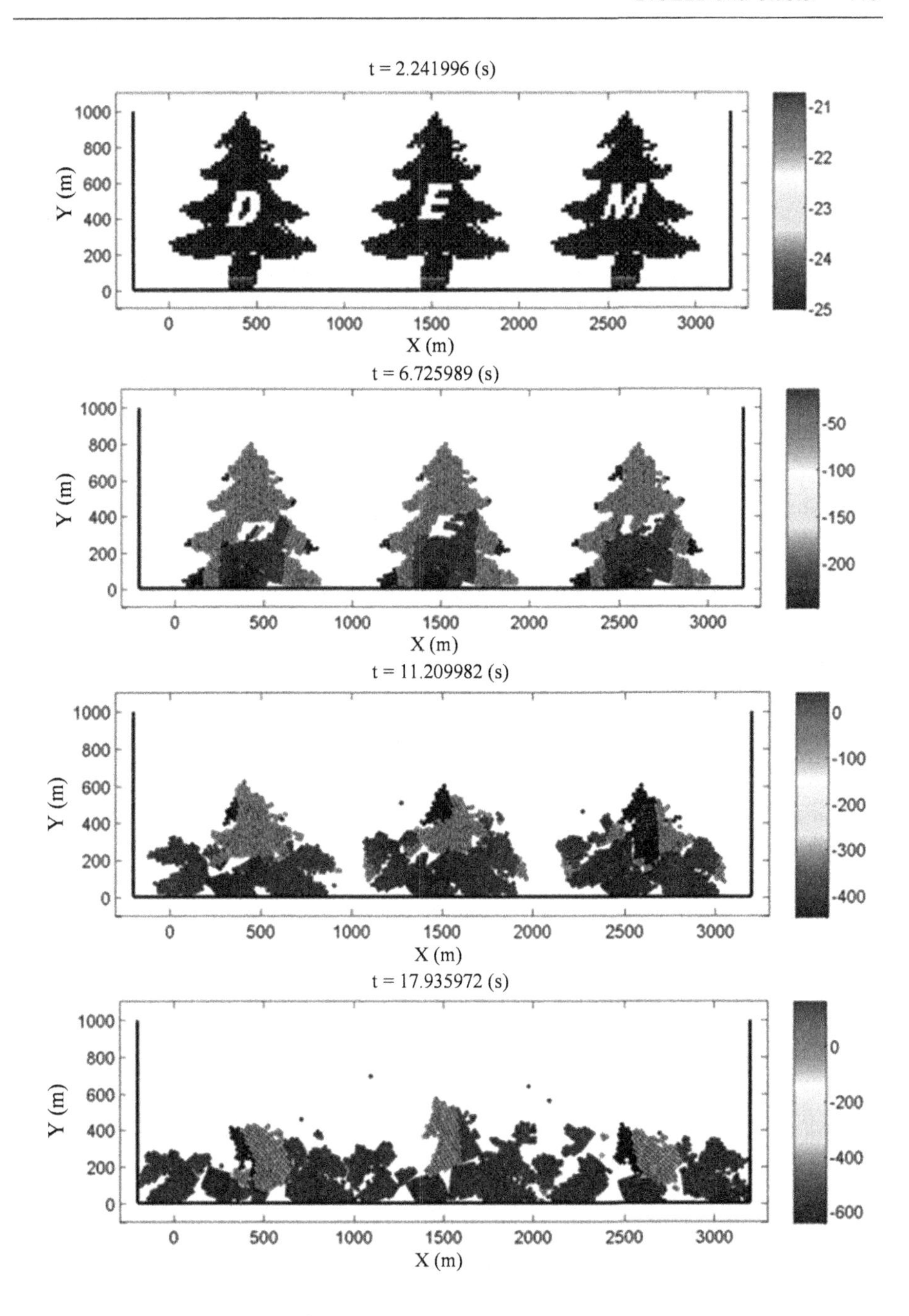

Figure 4.5. *Collapse process for the three trees under gravity using parallel DICE2D. For a color version of the figure, see www.iste.co.uk/zhao/computing.zip*

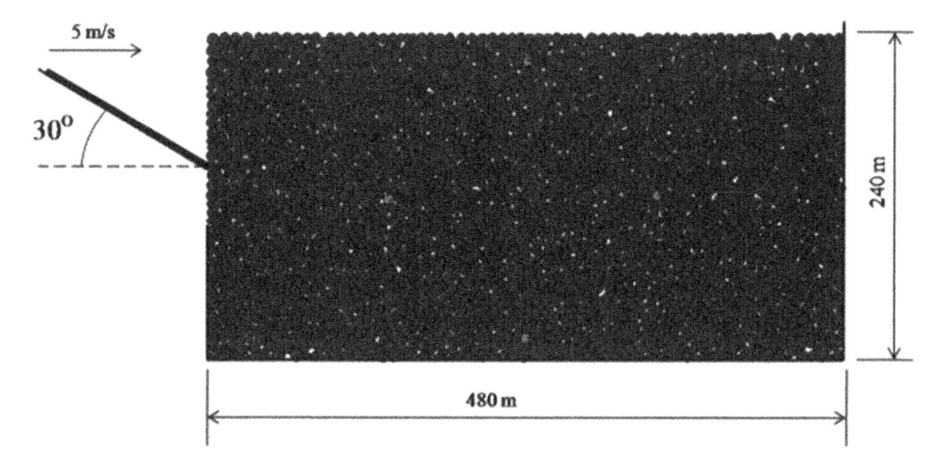

Figure 4.6. *Computational model for the rock cutting problem*

Number of particles	2,843	Normal viscous coefficient (s)	0.0
Mean particle size (m)	3.4	Friction angle	30
Density (kg/m^3)	1,000	WRT	0.1
Normal stiffness (N/m)	1e8	Local damping	0.01
Shear stiffness (N/m)	3e7	Time step reduction factor	0.1
Tension strength (N)	4e7	Total steps	10,000
Cohesion (N)	2e8	Gravitational acceleration (m/s^2)	10

Table 4.2. *Model parameters for the rock cutting problem*

Figure 4.7 shows the simulated rock cutting process. Fractures form before the cutter reaches the fracture front. After a chip is formed, other cracks will form (see Figure 4.7). By repeating the process, rock fragments are formed. The cutting depth, cutting velocity and cutting angle will influence the cutting profile, cutting force and cutting energy. Relationships between these variables can be obtained from a parameter analysis using the parallel DICE2D.

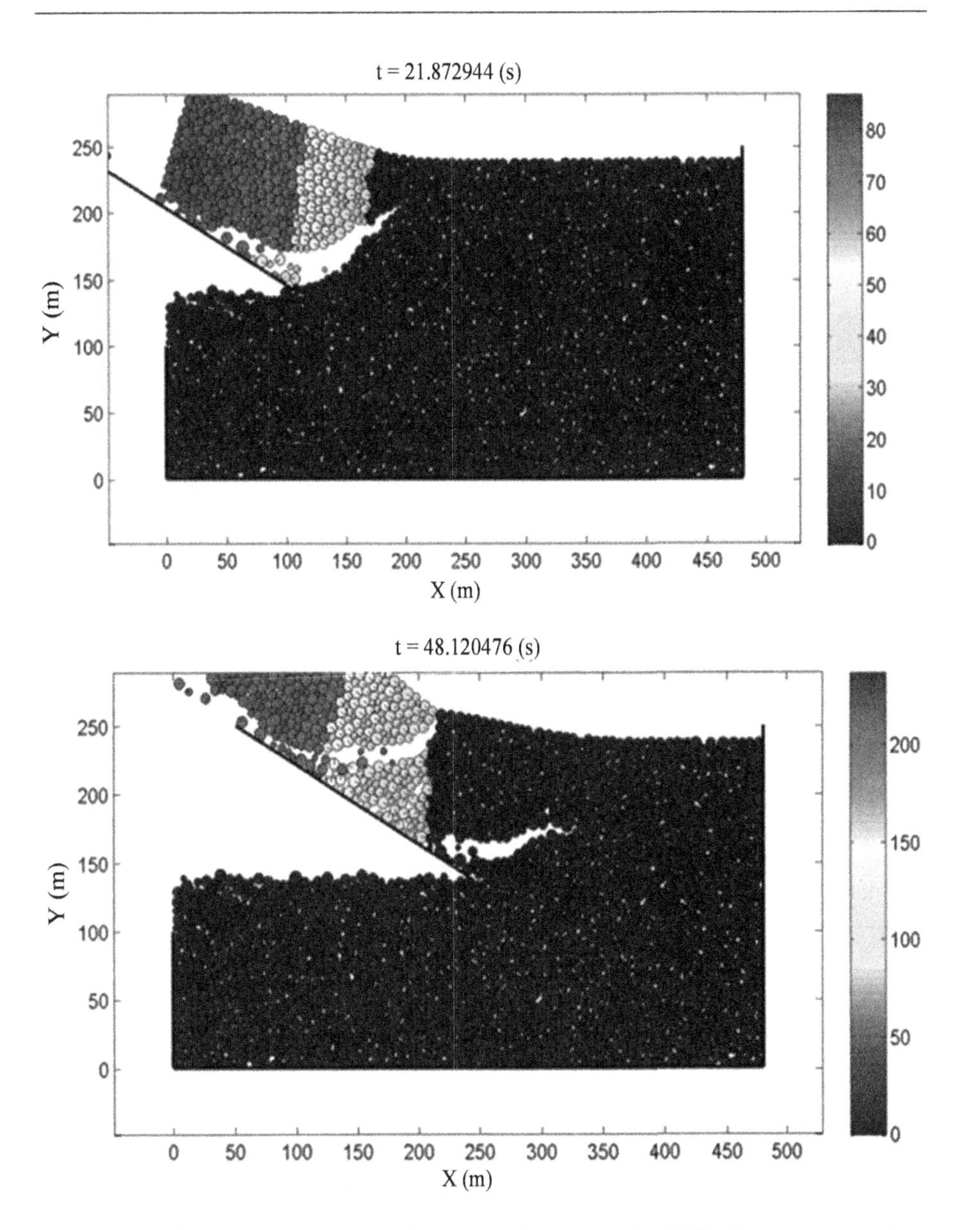

Figure 4.7. *Simulation results for the rock cutting using DICE2D. For a color version of the figure, see www.iste.co.uk/zhao/computing.zip*

4.3.3. *Slope stability analysis*

In this example, a rock slope stability problem is solved using DICE2D. The slope will fail under gravity because of external loading such as earthquakes or a change of the material properties due to environmental or human activities. Classical analysis methods, such as the limited equilibrium analysis [DUN 96], cannot provide for failure processes and failure patterns of the slope. These previous methods might result in overly conservative designs that consume more money than necessary, or, in the worst case, an underestimated safety factor (SOF) design would endanger the slope and potentially cause the loss of properties and lives. Two classical approaches are available for numerical simulations to obtain the failure process of a slope. In the first approach,the strength reduction method [DAW 99], the design material properties are inputted as initial parameters. The external loading is kept constant during calculation. Then, by decreasing the properties of the input material, the computational model will fail when a parameter (strength) is less than a certain value. The ratio between the critical value and the design value is calculated as the SOF. However, the strength reduction method is feasible only for computational models with single or a few adjustable mechanical parameters. The other approach is the gravity increase method [LI 09], which can be regarded as a numerical simulation of a centrifuge test. In this method, the design parameters of material are kept constant during calculation, whereas the gravitational acceleration increases. The model will fail under a certain gravitational acceleration, for example xg. Then, x is estimated as the SOF. The gravity increase method is useful for engineering projects with failure processes dominated by gravitational force.

In this section, a rock slope is simulated using DICE2D (Figure 4.8). The dimension of the slope is approximately 700 m × 400 m (the details are displayed in Figure 4.8). The slope comprises three parts: the base rock, the weak layer and the top slope block. A total of 9,129 particles are present in the model. The model parameters and mechanical properties of the corresponding materials are listed in Tables 4.3 and 4.4, respectively. The bottom, left and right of the model are fixed by three walls. A measuring point is established at the top surface to record the settlement history under different gravitational accelerations. Figures 4.9 and 4.10 show the modeling results for the rock slope under different gravitational accelerations. The slope fails under gravity. A small portion of the weak layer fails under $3g$. Then, a main fracture forms along the weak layer under $5g$ (see Figure 4.9). Finally, the slope will slide along the failure surface under $7g$ and $8g$ (see Figure 4.10). This result is a typical slope failure pattern. To obtain a more quantitative observation, the settlement history of the measured point is shown in Figure 4.11. The settlement is stable when the acceleration is less than $4g$. Therefore, the SOF of the slope is approximately 4.

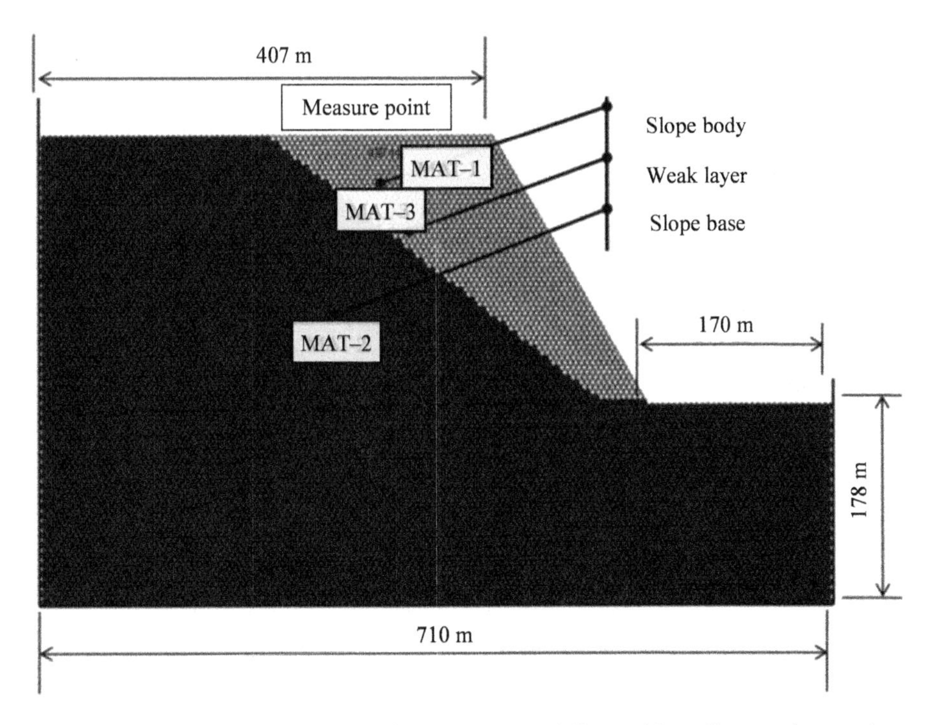

Figure 4.8. *Computational model for the slope stability problem. For a color version of the figure, see www.iste.co.uk/zhao/computing.zip*

Number of particles	10,951/10,966	Time step reduction factor	0.1
Mean particle size (m)	2.5	Total steps	40,000
Local damping	0.8	Gravitational acceleration (m/s^2)	10

Table 4.3. *Model parameters for the slope stability problem*

ID	Density (kg/m^3)	Normal stiffness (N/m)	Shear stiffness (N/m)	Tension strength (N)	Cohesion (N)	Friction angle (°)
1	2,000	1.0e9	0.3e9	3.2e8	2.4e8	60
2	2,000	1.0e9	0.3e9	3.2e8	2.4e8	60
3	2,000	3.0e8	0.9e8	0	0	10
4	3,000	1.0e10	3.0e9	2e10	2.4e8	30

Table 4.4. *Material parameters for the slope stability problem*

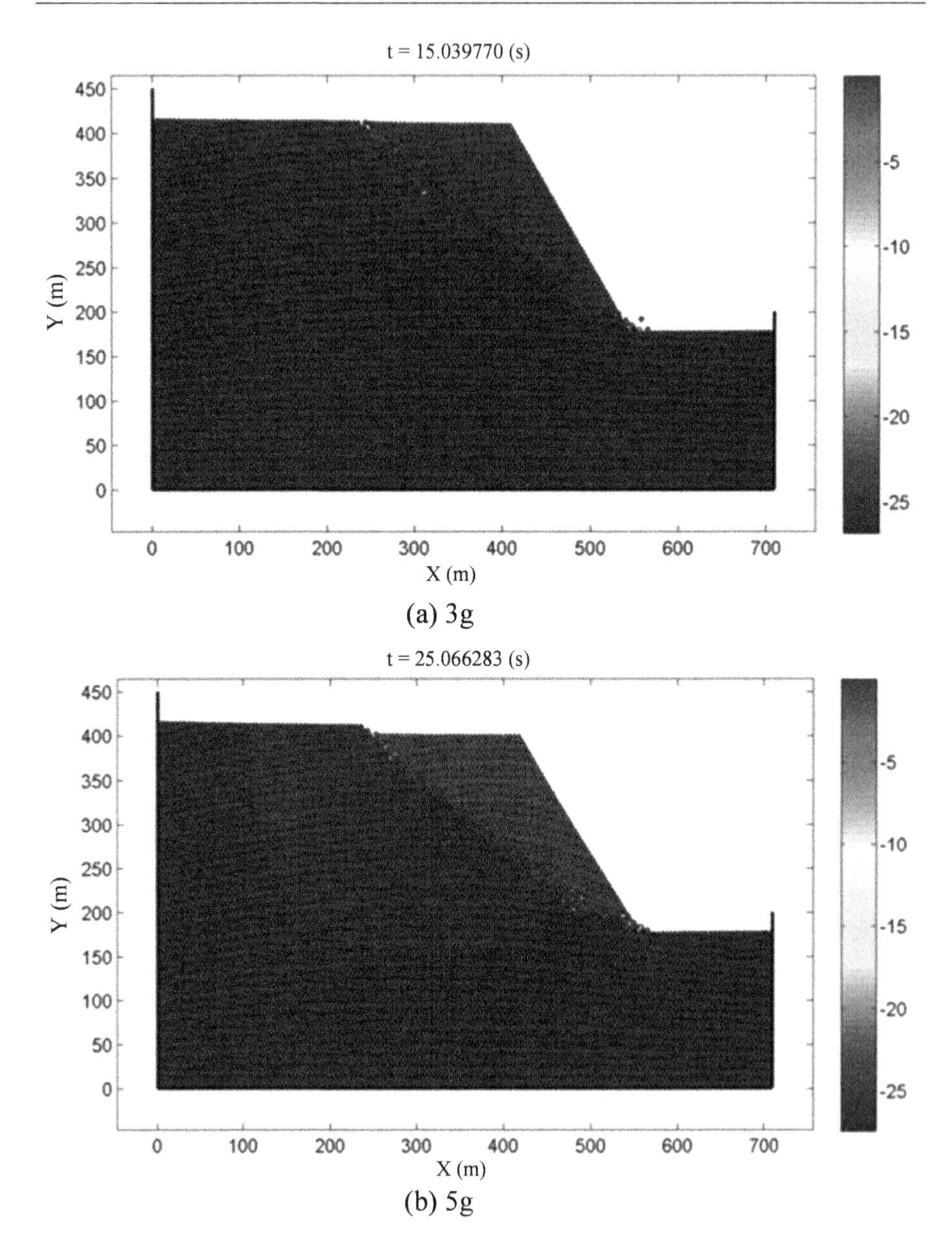

Figure 4.9. *Slope configurations under different gravitational accelerations (3g and 5g). For a color version of the figure, see www.iste.co.uk/zhao/computing.zip*

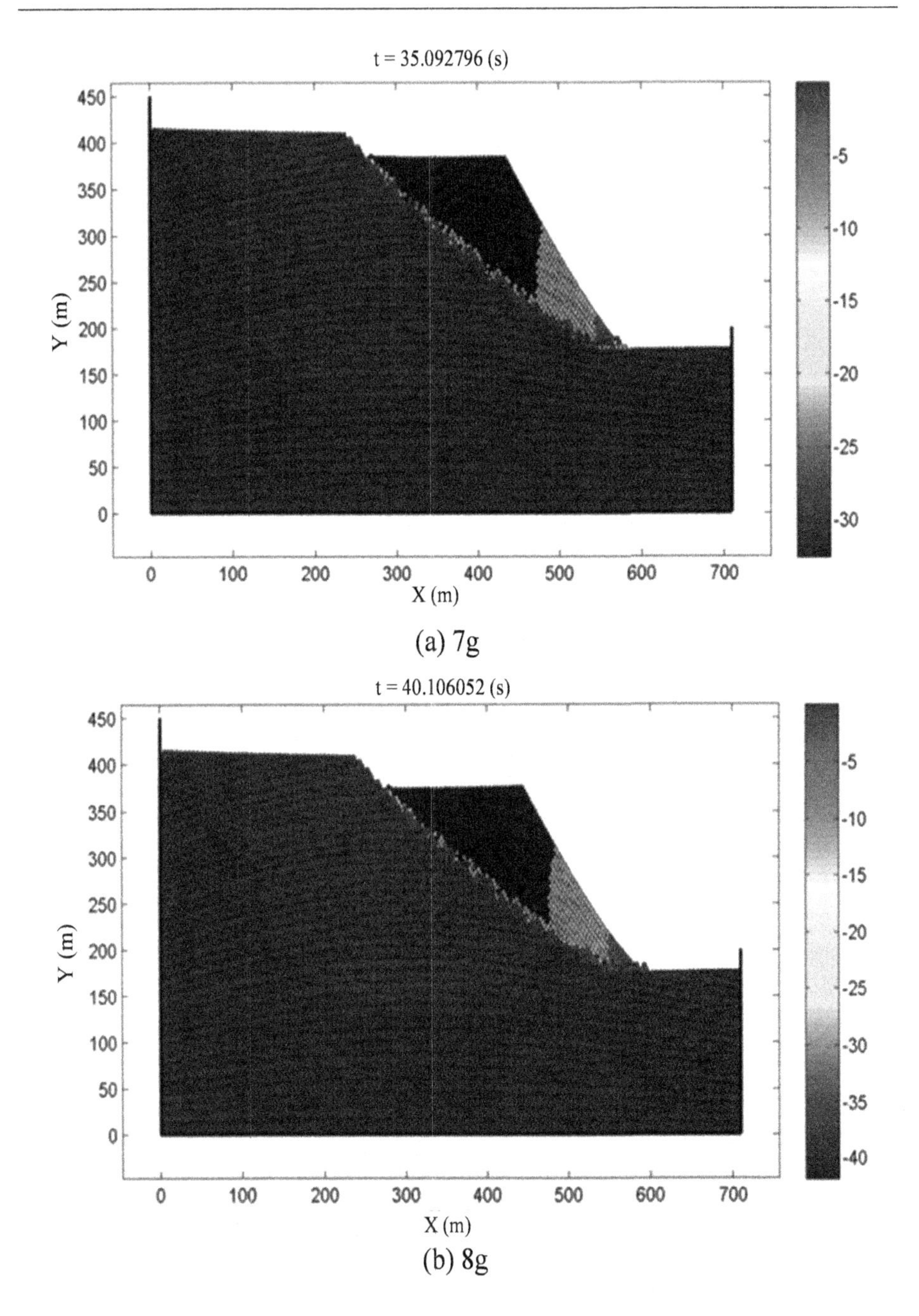

Figure 4.10. *Slope configurations under different gravitational accelerations (7g and 8g). For a color version of the figure, see www.iste.co.uk/zhao/computing.zip*

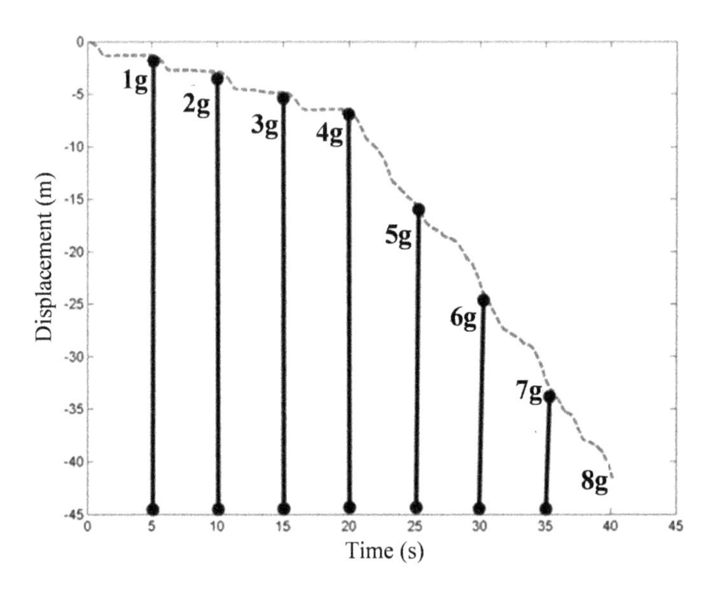

Figure 4.11. *History of the displacement in the y-direction of the measured point of the slope*

In real engineering applications, one common method used to reinforce a slope is installing rock bolts. As shown in Figure 4.12, the slope model in Figure 4.8 is enriched using rock bolts. The model parameters and material parameters are taken to be identical as the previous model. The mechanical properties of the rock bolt are adopted as a material with ID 4. A higher strength and elastic modulus are used (see Table 4.4). The rock bolt is simulated using a group of bonded particles. The detachment and sliding between the rock bolt and rock are considered through interactions between particles of the rock and rock bolt. Because the rock bolt is approximated using a particle model, this method is only a preliminary analysis. The main purpose of this method is to show the ability of DICE2D to model reinforced-slope stability problem. Figure 4.13 shows the history of settlement at the measured point under different gravitational accelerations. A stable solution is obtained for the reinforced slope even under 8*g*. Moreover, the displacement contours of the slope under the corresponding gravitational accelerations are shown in Figures 4.14 and 4.15. The slope remains intact under 8*g*; only some local failures are noted. Therefore, the rock bolt reinforced slope is safer than the original slope. In addition, rock bolting is a useful method for reinforcing the rock slope.

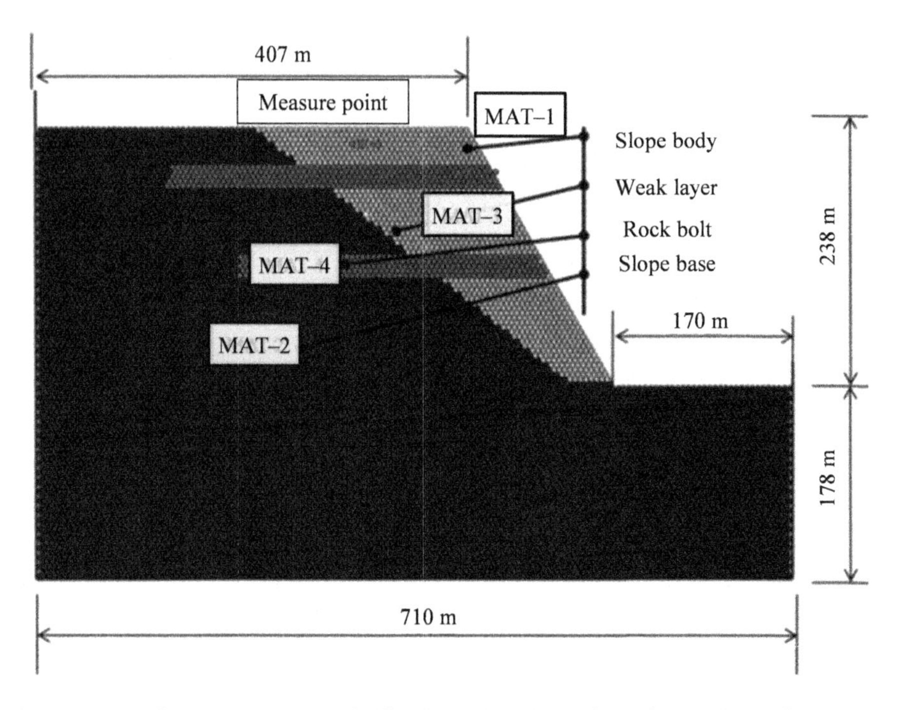

Figure 4.12. *Computational model for the rock bolt reinforced rock slope. For a color version of the figure, see www.iste.co.uk/zhao/computing.zip*

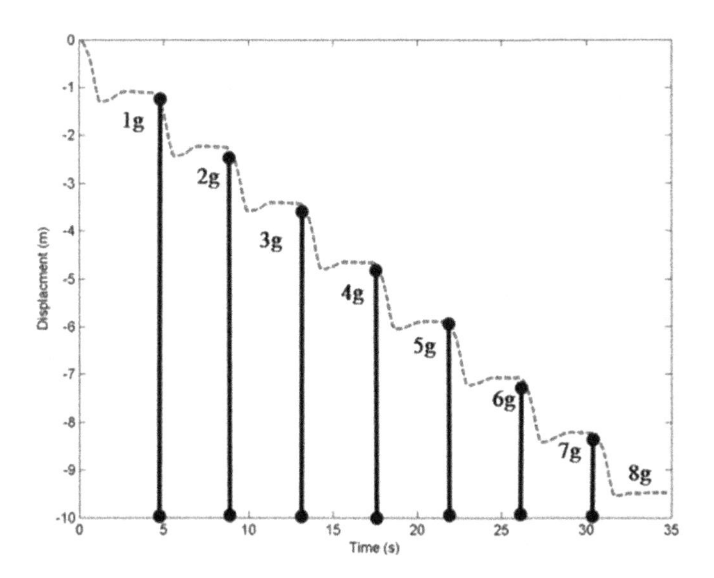

Figure 4.13. *History of the displacement in the y-direction of the reinforced slope*

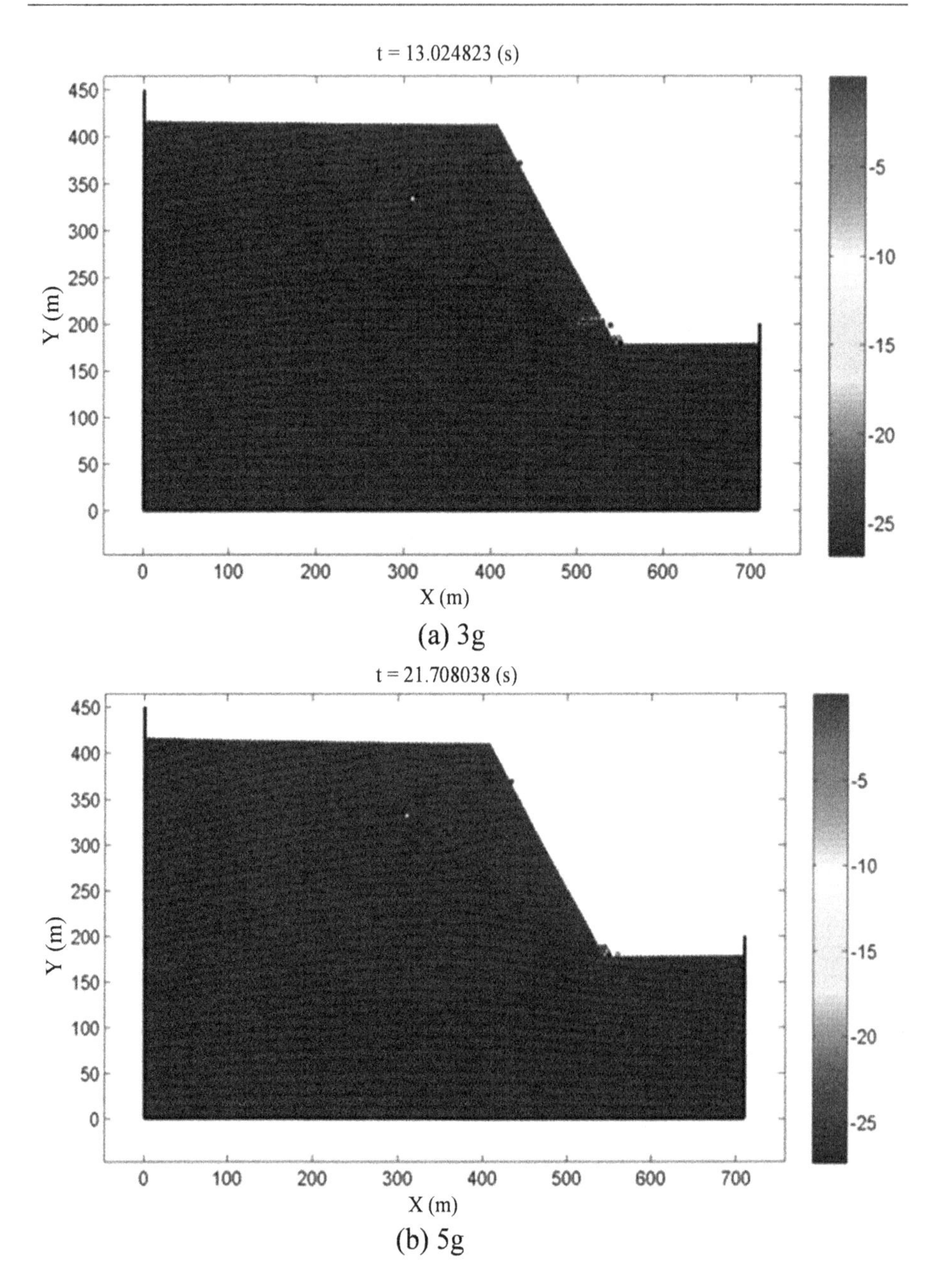

Figure 4.14. *Configurations of a slope with rock bolts under different gravitational accelerations (3g and 5g). For a color version of the figure, see www.iste.co.uk/zhao/computing.zip*

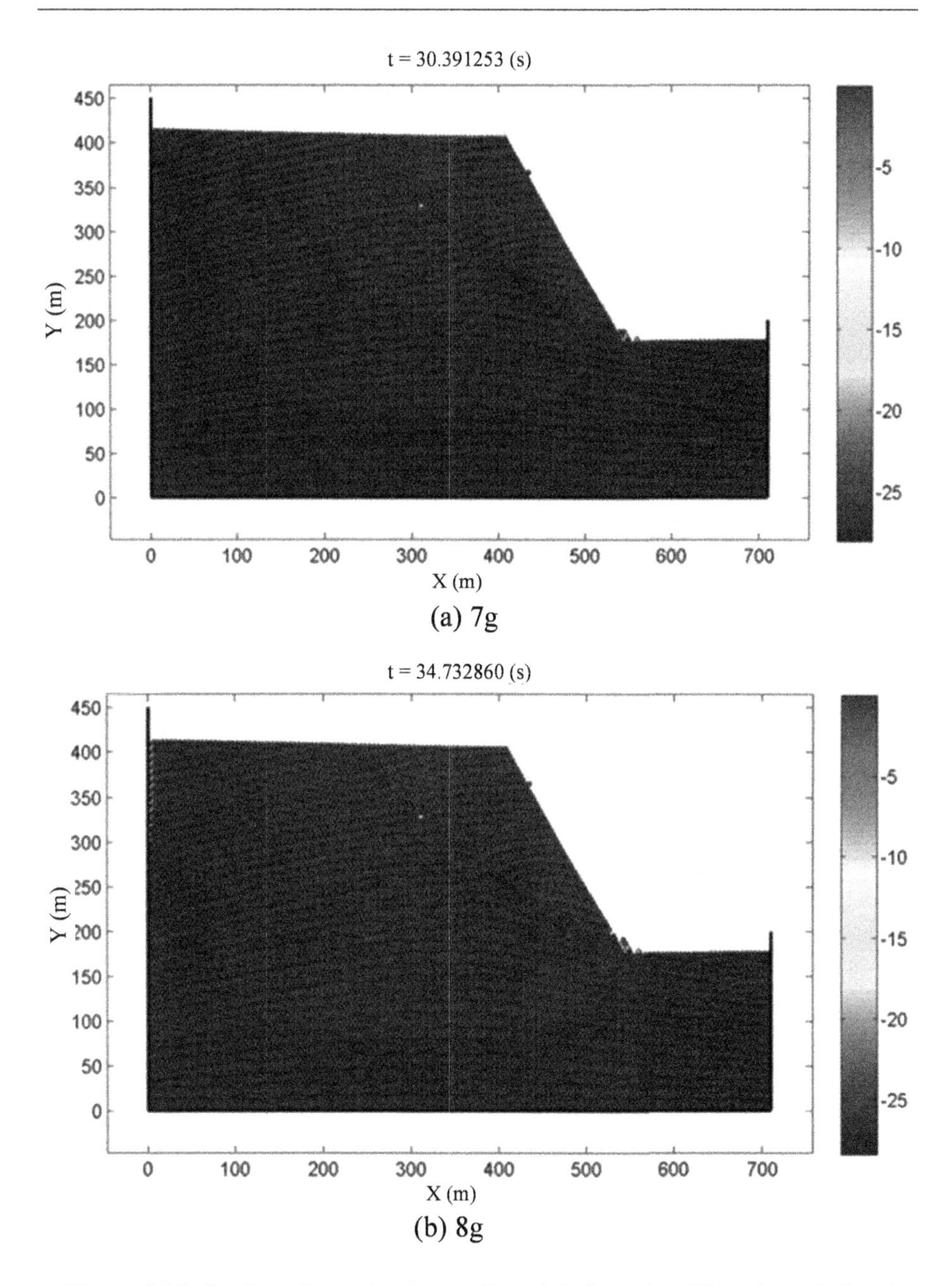

Figure 4.15. *Configurations of a slope with rock bolts under different gravitational accelerations (7g and 8g). For a color version of the figure, see www.iste.co.uk/zhao/computing.zip*

4.3.4. *Interaction between ground structure and underground structure*

In this example, a ground structure and underground structure interaction problem is simulated using DICE2D. The computational model is shown in Figure 4.16. A megaframe structure is built on the base rock; the base comprises two intact blocks with a weak fault. To investigate the influence of the underground structure, another computational model with an underground cavern is constructed (Figure 4.17). A large cavern is excavated just below the megaframe structure. The digital image modeling technique was adopted in this example as well. The model and material parameters are listed in Tables 4.5 and 4.6, respectively.

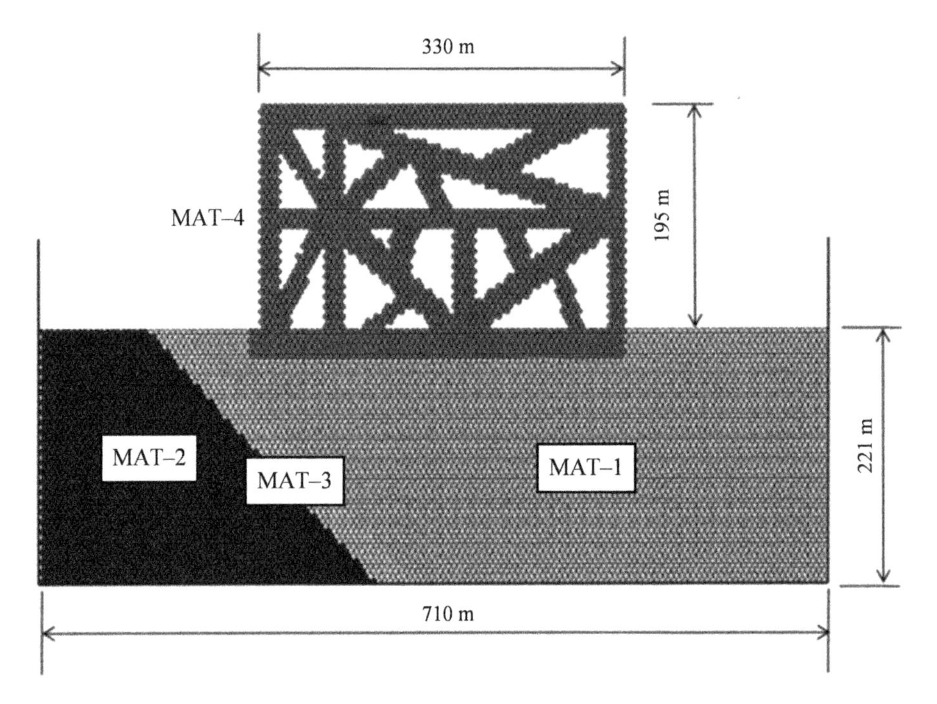

Figure 4.16. *Computational model of the interaction problem without an underground cavern. For a color version of the figure, see www.iste.co.uk/zhao/computing.zip*

Number of particles	9,129	Time step reduction factor	0.1
Mean particle size (m)	2.5	Total steps	40,000
Local damping	0.8	Gravitational acceleration (m/s^2)	10

Table 4.5. *Model parameters for the structure–tunnel interaction problem*

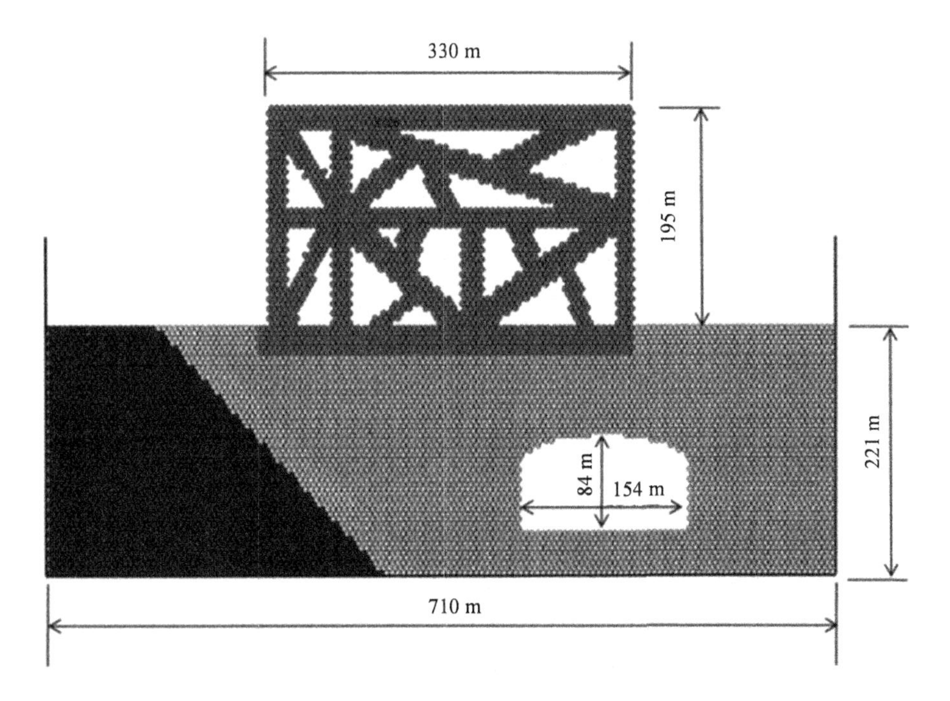

Figure 4.17. *Computational model of the interaction problem with an underground cavern. For a color version of the figure, see www.iste.co.uk/zhao/computing.zip*

ID	Density (kg/m^3)	Normal stiffness (N/m)	Shear stiffness (N/m)	Tension strength (N)	Cohesion (N)	Friction angle (°)
1	1,000	4.00e8	1.20e8	6e8	7.2e6	30
2	1,000	3.20e8	0.96e8	1.2e8	7.2e6	30
3	1,000	2.00e8	0.60e8	3.6e6	7.2e6	10
4	1,000	4.00e9	1.20e8	1.2e13	7.2e7	30

Table 4.6. *Material parameters for the structure–tunnel interaction problem*

The gravity increase method (numerical centrifuge test) was used to obtain the failure pattern and the SOF of these two models. The simulation results of the ground structure without a cavern are shown in Figures 4.18 and 4.19. When the gravitational acceleration is 1*g*, the settlement along the *y*-direction is stable (see Figure 4.18(a)). The biggest displacement occurs at the upper right part of the megaframe structure. As shown in Figure 4.18(a) and (b), the displacement distribution pattern does not change under gravitational accelerations of $2g$ and $3g$. When the gravitational acceleration is $4g$, both the structure and the base will fracture (see Figure 4.19(b)). The building is broken around the upper right part (the maximum deformation zone). The interface between the fault and the base rock is also damaged. The SOF of the computational model is estimated to be approximately 3.0.

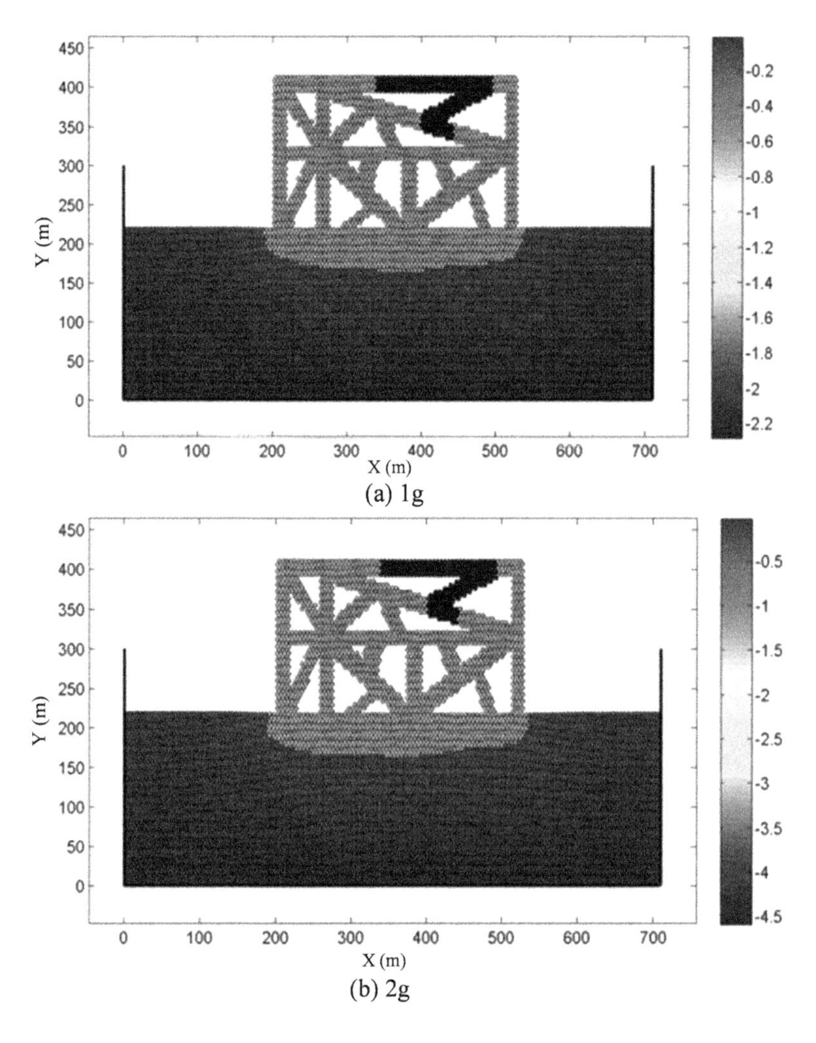

Figure 4.18. *Displacement in the y-direction of the model under gravitational accelerations of 1g and 2g. For a color version of the figure, see www.iste.co.uk/zhao/computing.zip*

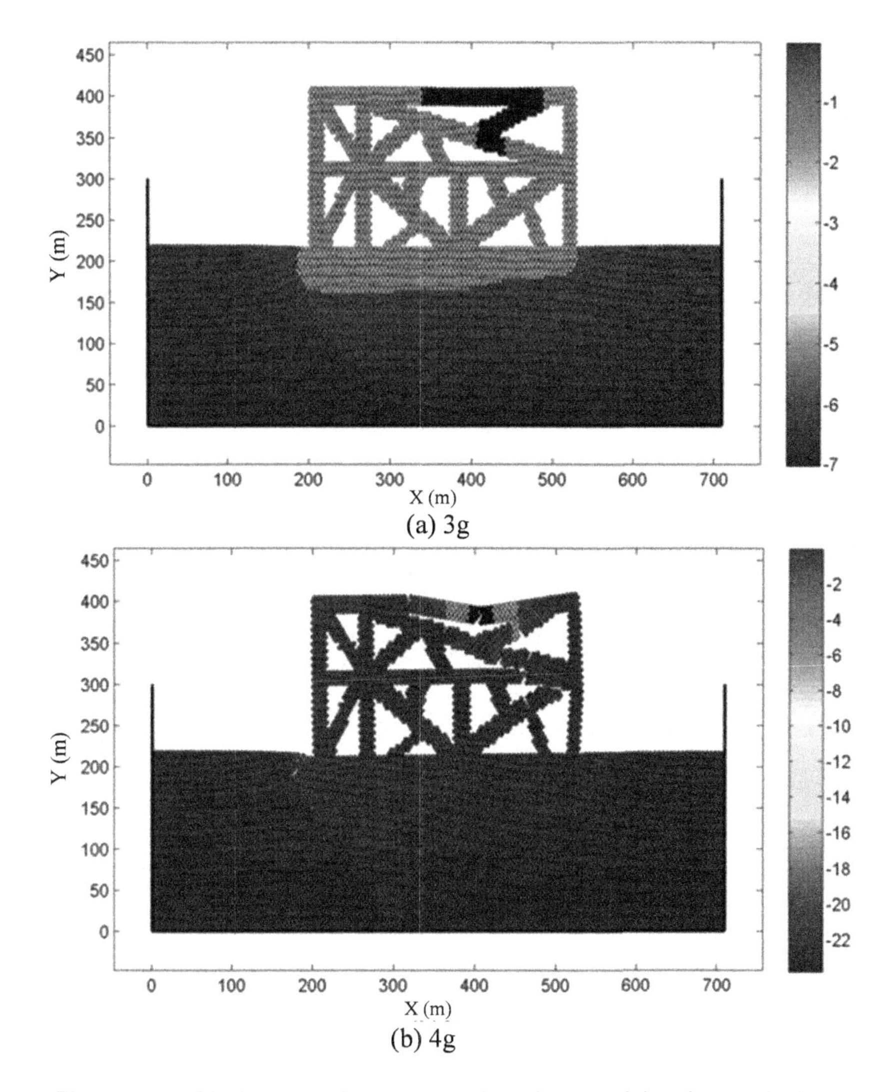

Figure 4.19. *Displacement in the y-direction of the model under gravitational accelerations of 3g and 4g. For a color version of the figure, see www.iste.co.uk/zhao/computing.zip*

The simulation results of the computational model with an underground cavern are shown in Figures 4.20 and 4.21. The model is stable under a gravitational acceleration of 1g. The displacement distribution in the y-direction is shown in Figure 4.20(a). The main settlement happens in the right-hand corner of the ground structure and the upper part of the cavern. The cavern collapses under a gravitational acceleration of 2g (see Figure 4.20(b)). When the gravitational acceleration is 3g, the cavern will completely fail, and fractures reach the upper ground. The base of the structure is also influenced. Both the ground structure and the cavern are completely broken under gravitational acceleration of 4g (Figure 4.21(b)). From the modeling results, the SOF of the second model is approximately 1. Therefore, the underground excavation can substantially affect the safety of a ground structure, which must be considered carefully during applications.

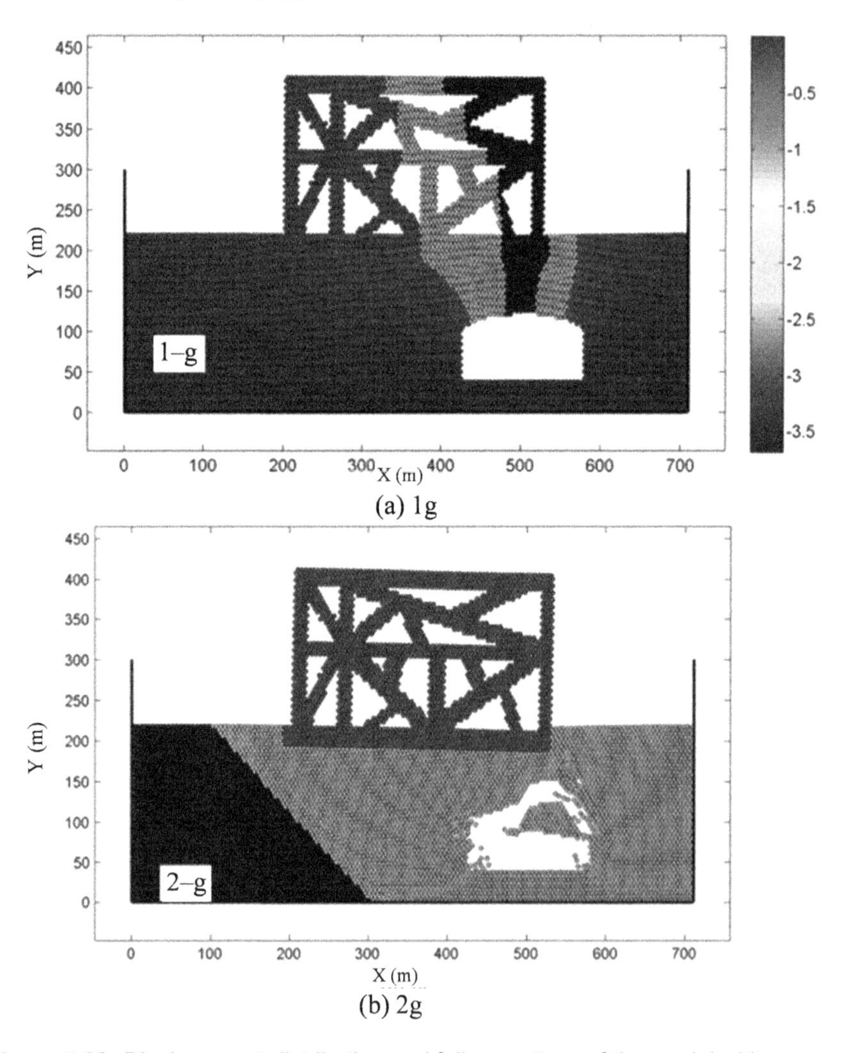

Figure 4.20. *Displacement distribution and failure pattern of the model with a cavern under gravitational accelerations of 1g and 2g, respectively. For a color version of the figure, see www.iste.co.uk/zhao/computing.zip*

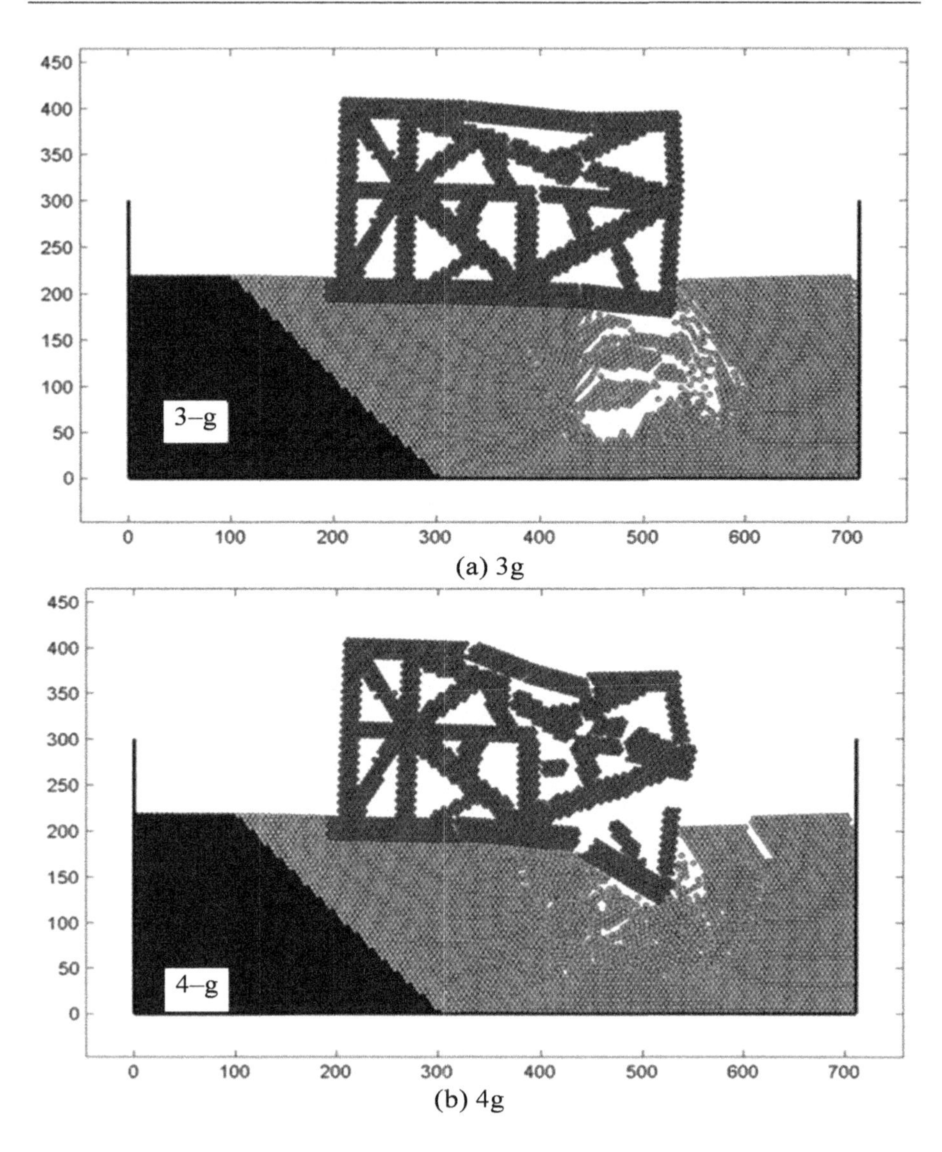

Figure 4.21. *Displacement distribution and failure pattern of the model with a cavern under gravitational accelerations of 3g and 4g, respectively. For a color version of the figure, see www.iste.co.uk/zhao/computing.zip*

4.4. Conclusion

In this chapter, the parallel DICE2D was run on a middle-sized cluster. The parallel DICE2D developed in the MATLAB environment can directly use the computational resources on the cluster. The parallel DICE2D shows a better performance than the GPU DICE2D when run on a cluster. The parallel DICE2D is also used to solve some rock engineering problems. The trees collapse problem considers large deformations, fragmentations and irregular particle interactions. The rock cutting problem shows the ability of DICE2D to model multiple crack interactions and the fragmentation of rock. The slope stability problem and ground structure interaction problem are also solved using the parallel DICE2D. Both realistic failure patterns and SOFs are obtained.

Bibliography

[ABE 04] ABE S., PLACE D., MORA P., "A parallel implementation of the lattice solid model for the simulation of rock mechanics and earthquake dynamics", *Pure and Applied Geophysics*, vol. 16, no. 11-12, pp. 2265-2277, 2004.

[AND 08] ANDERSON J.A., LORENZ C.D., TRAVESSET A., "General purpose molecular dynamics simulations fully implemented on graphics processing units", *Journal of Computational Physics*, vol. 227, no. 10, pp. 5342-5359, 2008.

[BRO 08] BROWN E.T., *Estimating the Mechanical Properties of Rock Masses*, SHIRMS, 2008.

[CAM 13] CAMONES L.A.M., VARGAS E.D.A., DE FIGUEIREDO R.P., *et al.*, "Application of the discrete element method for modeling of rock crack propagation and coalescence in the step-path failure mechanism", *Engineering Geology*, vol. 153, no. 8, pp. 80-94, 2013.

[CHA 06] CHANG V., "Experiments and investigations for the personal high performance computing (HPC) built on top of the 64-bit processing and clustering systems", *13th Annual IEEE International Symposium*, Germany, pp. 27-30, 2006.

[CHO 07] CHO N., MARTIN C.D., SEGO D.C., "A clumped particle model for rock", *International Journal of Rock Mechanics and Mining Sciences*, vol. 44, pp. 997-1010, 2007.

[CUN 71] CUNDALL P.A., "A computer model for simulating progressive, large-scale movements in blocky rock systems", *Proceedings of the Symposium of International Society of Rock Mechanics*, vol. 2, no. 8, pp. 2-8, 1971.

[CUN 01] CUNDALL P.A., "A discontinuous future for numerical modelling in geomechanics?", *Proceedings of the Institution of Civil Engineers-Geotechnical Engineering*, vol. 149, no. 1, pp. 41-47, 2001.

[DAW 99] DAWSON E.M., ROTH W.H., DRESCHER A., "Slope stability analysis by strength reduction", *Geotechnique*, vol. 49, no. 6, pp. 835-840, 1999.

[DIC 96] DICK B., JACQUELINE F., BRADFORD N., *pThreads Programming*, O'Reilly Media, 1996.

[DIC 11] DICK C., GEORGII J., WESTERMANN R., "A real-time multigrid finite hexahedra method for elasticity simulation using CUDA", *Simulation Modelling Practice and Theory*, vol. 19, no. 2, pp. 801-816, 2011.

[DOW 99] DOWDING C.H., DMYTRYSHYN O., BELYTSCHKO T.B., "Parallel processing for a discrete element program", *Computers and Geotechnics*, vol. 25, no. 4, pp. 281-285, 1999.

[DUN 96] DUNCAN J.M., "State of the art, limit equilibrium and finite-element analysis of slopes", *Journal of Geotechnical Engineering*, vol. 122, no. 7, pp. 577-596, 1996.

[ELS 08] ELSEN E., LEGRESLEY P., DARVE E., "Large calculation of the flow over a hypersonic vehicle using a GPU", *Journal of Computational Physics*, vol. 227, no. 24, pp. 10148-10161, 2008.

[FAN 11] FANG H.F., TADE M.O., LI Q., "A numerical study on the role of geometry confinement and fluid flow in colloidal self-assembly", *Powder Technology*, vol. 214, no. 3, pp. 283-291, 2011.

[GOP 13] GOPALAKRISHNAN P., TAFTI D., "Development of parallel DEM for the open source code MFIX", *Powder Technology*, vol. 235, pp. 33-41, 2013.

[HER 11] HERRMANN F., SILBERHOLZ J., TIGLIO M., "Black hole simulations with CUDA", in HWU W.-M.W., *GPU Computing Gems*, Elsevier, pp. 103-111, 2011.

[HOR 11] HORI C., GOTOH H., IKARI H., *et al.*, "GPU-acceleration for moving particle semi-implicit method", *Computers and Fluids*, vol. 51, no. 1, pp. 174-183, 2011.

[HRE 41] HRENNIKOFF A., "Solution of problems of elasticity by the framework method", *ASME Journal of Applied Mechanics*, vol. 8, pp. A619-A715, 1941.

[HUG 80] HUGHES T.J.R., WINGET J., "Finite rotation effects in numerical integration of rate constitutive equations in large deformation analysis", *International Journal of Numerical Methods in Engineering*, vol. 15, pp. 1862-1867, 1980.

[ITA 08] ITASCA, PFC2D (Particle Flow Code in 2 Dimensions) Version 4.0, IGC, Minneapolis, MN, 2008.

[JIA 05] JIANG M.J., YU H.S., HARRIS D., "A novel discrete model for granular material incorporating rolling resistance", *Computers and Geotechnics*, vol. 32, no. 5, pp. 340-357, 2005.

[JOL 10] JOLDES G.R., WITTEK A., MILLER K., "Real-time nonlinear finite element computations on GPU – Application to neuro surgical simulation", *Computer Methods in Applied Mechanics and Engineering*, vol. 199, no. 49-52, pp. 3305-3314, 2010.

[KAC 10] KAČIANAUSKAS R., MAKNICKAS A., KAČENIAUSKAS A., *et al.*, "Parallel discrete element simulation of poly-dispersed granular material", *Advances in Engineering Software*, vol. 41, no. 1, pp. 52-63, 2010.

[KAZ 13] KAZERANI T., "Effect of micromechanical parameters of microstructure on compressive and tensile failure process of rock", *International Journal of Rock Mechanics and Mining Sciences*, vol. 64, pp. 44-55, 2013.

[KAZ 10] KAZERANI T., ZHAO J. "Micromechanical parameters in bonded particle method FOR modelling of brittle material failure", *International Journal for Numerical and Analytical Methods in Geomechanics*, vol. 34, no. 18, pp. 1877-1895, 2010.

[KOM 10] KOMATITSCH D., ERLEBACHER G., GÖDDEKE D., *et al.*, "High-order finite-element seismic wave propagation modeling with MPI on a large GPU cluster", *Journal of Computational Physics*, vol. 229, pp. 7692-7714, 2010.

[LAU 00] LAUBSCHER D., *A Practical Guide Manual on Block Caving*, ICS, 2000.

[LI 09] LI L.C., TANG C.A., ZHU W.C., *et al.*, "Numerical analysis of slope stability based on the gravity increase method", *Computer and Geotechnics*, vol. 36, no. 7, pp. 1246-1258, 2009.

[LI 13] LI D., WONG L.N.Y., "The Brazilian disc test for rock mechanics applications: review and new insights", *Rock Mechanics and Rock Engineering*, vol. 46, pp. 269-287, 2013.

[LIS 14] LISJAK A., GRASSELLI G., "A review of discrete modeling techniques for fracturing processes in discontinuous rock masses", *Journal of Rock Mechanics and Geotechnical Engineering*, vol. 6, no. 4, pp. 301-314, 2014.

[LIU 03] LIU S.H., SUN D.A., WANG Y.S., "Numerical study of soil collapse behavior by discrete element modelling", *Computers and Geotechnics*, vol. 30, no. 5, pp. 399-408, 2003.

[MA 11] MA Z., FENG C., LIU T., *et al.*, "A GPU accelerated continuous-based discrete element method for elasto dynamics analysis", *Advanced Materials Research*, vol. 320, pp. 329-334, 2011.

[MAK 06] MAKNIČKAS A., KAČENIAUSKAS A., KAČIANAUSKAS R., *et al.*, "Parallel DEM software for simulation of granular media", *Informatica*, vol. 17, no. 2, pp. 207-224, 2006.

[MAR 11] MARKAUSKAS D., KAČENIAUSKAS A., MAKNICKAS A., "Dynamic domain decomposition applied to Hopper discharge simulation by discrete element method", *Information Technology and Control*, vol. 40, no. 4, pp. 286-292, 2011.

[MCN 93] MCNEARNY R.L., ABEL JR. J.F., "Large-scale two-dimensional block caving model tests", *International Journal of Rock Mechanics and Mining Sciences*, vol. 30, no. 2, pp. 93-109, 1993.

[MUN 95] MUNJIZA A., OWEN D.R.J., BICANIC N., "A suite element – discrete element approach to the simulation of rode blasting problems", *Engineering Computations*, vol. 12, pp. 145-174, 1995.

[MUN 04] MUNJIZA A., *The Combined Finite-Discrete Element Method*, John Wiley & Sons, 2004.

[NIS 11] NISHIURA D., SAKAGUCHI H., "Parallel-vector algorithms for particle simulations on shared-memory multiprocessors", *Journal of Computational Physics*, vol. 230, no. 5, pp. 1923-1938, 2011.

[OPE 10] OPENMP, OpenMP News, available at http://www.openmp.org, 2010.

[ORT 10] ORTEGA L., RUEDA A., "Parallel drainage network computation on CUDA", *Computers and Geosciences*, vol. 36, pp. 171-178, 2010.

[POD 11] PODLICH N.C., ABBO A.J., SLOAN S.W., "Application of GPU computing to upper bound rigid block analysis", in KHALILI N., OESER M. (eds), *Proceedings of the 13th International Conference of IACMAG*, Melbourne, Australia, pp. 60-65, 2011.

[POT 04] POTYONDY D.O., CUNDALL P.A., "A bonded-particle model for rock", *International Journal of Rock Mechanics and Mining Sciences*, vol. 41, no. 8, pp. 1329-1364, 2004.

[RIC 08] RICK M., *CPU Designer Debate Multi-core Future*", EE Times, 2008.

[SAI 10] SAINIO J., "CUDAEASY – a GPU accelerated cosmological lattice program", *Computer Physics Communications*, vol. 181, pp. 906-912, 2010.

[SCH 04] SCHAFER B.C., QUIGLEY S.F., CHAN A.H.C., "Acceleration of the discrete element method (DEM) on a reconfigurable co-processor", *Computers and Structures*, vol. 82, no. 20-21, pp. 1707-1718, 2004.

[SCH 13] SCHOLTES L., DONZE F.-V., "A DEM model for soft and hard rocks: role of grain interlocking on strength", *Journal of the Mechanics and Physics of Solids*, vol. 61, no. 2, pp. 352-369, 2013.

[SIM 92] SIMO J.C., LAURSEN T.A., "An augmented Lagrangian treatment of contact problems involving friction", *Computers and Structures*, vol. 42, no. 1, pp. 97-116, 1992.

[STO 10] STONE J.E., HARDY D.J., UFIMTSEV I.S., *et al.*, "GPU-accelerated molecular modeling coming of age", *Journal of Molecular Graphics and Modelling*, vol. 29, pp. 116-125, 2010.

[TAK 09] TAKAHASHI T., HAMADA T., "GPU-accelerated boundary element method for Helmholtz' equation in three dimensions", *International Journal for Numerical Methods in Engineering*, vol. 80, no. 10, pp.1295-1321, 2009.

[TBB 10] TBB, Threading Building Blocks (Intel TBB), available at http://www.threading buildingblocks.org/, 2010.

[TRU 08] TRUEMAN R., CASTRO R., HALIM A., "Study of multiple draw-zone interaction in block caving mines by means of a large 3D physical model", *International Journal of Rock Mechanics and Mining Sciences*, vol. 45, no. 7, pp. 1044-1051, 2008.

[WAL 09] WALSH S.D.C., SAAR M.O., BAILEY P., *et al.*, "Accelerating geoscience and engineering system simulations on graphics hardware", *Computers and Geosciences*, vol. 35, no. 12, pp. 2353-2364, 2009.

[WAL 09] WALTHER J.H., SBALZARINI I.F., "Large-scale parallel discrete element simulations of granular flow", *Engineering Computations*, vol. 26, no. 6, pp. 688-697, 2009.

[WAN 13] WANG L.X., LI S.H., ZHANG G.X., *et al.*, "A GPU-based parallel procedure for nonlinear analysis of complex structures using a coupled FEM/DEM approach", *Mathematical Problems in Engineering*, vol. 2013, Article ID 618980, 2013.

[XU 11] XU J., QI H., FANG X., *et al.*, "Quasi-real-time simulation of rotating drum using discrete element method with parallel GPU computing", *Particuolog*, vol. 9, pp. 446-450, 2011.

[YOO 07] YOON J., "Application of experimental design and optimization to PFC model calibration in uniaxial compression simulation", *International Journal of Rock Mechanics and Mining Sciences*, vol. 44, no. 6, pp. 871-889, 2007.

[ZHA 13a] ZHANG L., QUIGLEY S.F., CHAN A.H.C., "A fast scalable implementation of the two-dimensional triangular discrete element method on a GPU platform", *Advances in Engineering Software*, vol. 60-61, pp. 70-80, 2013.

[ZHA 13b] ZHAO G.F., FANG J.N., SUN L., *et al.*, "Parallelization of the distinct lattice spring model", *International Journal for Numerical and Analytical Methods in Geomechanics*, vol. 37, no. 1, pp. 51-74, 2013.

[ZHA 12] ZHAO G.F., KHALILI N., "Graphics processing unit based parallelization of the distinct lattice spring model", *Computers and Geotechnics*, vol. 42, pp. 109-117, 2012.

[ZHA 14] ZHAO G.F., RUSSELL A.R., ZHAO X.B., *et al.*, "Strain rate dependency of uniaxial tensile strength in Gosford sandstone by the Distinct Lattice Spring Model with X-ray micro CT", *International Journal of Solids and Structures*, vol. 51, no. 7-8, pp. 1587-1600, 2014.

[ZHA 08] ZHAO X.B., ZHAO J., CAI J.G., *et al.*, "UDEC modelling on wave propagation across fractured rock masses", *Computers and Geotechnics*, vol. 35, no. 1, pp. 97-104, 2008.

[ZHE 12] ZHENG J.W., AN X.H., HUANG M.S., "GPU-based parallel algorithm for particle contact detection and its application in self-compacting concrete flow simulations", *Computers and Structures*, vol. 112, pp. 193-204, 2012.

[ZHU 08] ZHU H.P., ZHOU Z.Y., YANG R.Y., *et al.*, "Discrete particle simulation of particulate systems: a review of major applications and findings", *Chemical Engineering Science*, vol. 63, no. 23, pp. 5728-5770, 2008.

[ZSA 09] ZSAKI A.M., "Parallel generation of initial element assemblies for two-dimensional discrete element simulations", *International Journal for Numerical and Analytical Methods in Geomechanics*, vol. 33, no. 3, pp. 377-389, 2009.

Index

H

I

L

M

N, P

Q

R

S

Printed in the United States
By Bookmasters